Describing the Hand of God

Describing the Hand of God

Divine Agency and Augustinian Obstacles to
the Dialogue between Theology and Science

Robert Brennan

PICKWICK *Publications* · Eugene, Oregon

DESCRIBING THE HAND OF GOD
Divine Agency and Augustinian Obstacles to the Dialogue between
Theology and Science

Copyright © 2015 Robert Brennan. All rights reserved. Except for brief quotations in critical publications or reviews, no part of this book may be reproduced in any manner without prior written permission from the publisher. Write: Permissions. Wipf and Stock Publishers, 199 W. 8th Ave., Suite 3, Eugene, OR 97401.

Pickwick Publications
An Imprint of Wipf and Stock Publishers
199 W. 8th Ave., Suite 3
Eugene, OR 97401

www.wipfandstock.com

ISBN 13: 978-1-62564-913-3

Cataloguing-in-Publication Data

Brennan, Robert

　　Describing the hand of God: divine agency and Augustinian obstacles to the dialogues between theology and science / Robert Brennan

　　xvi + 288 pp. ; 23 cm. Includes bibliographical references.

　　ISBN 13: 978-1-62564-913-3

　　1. Religion and Science 2. Darwin, Charles, 1809–1882 3. Newton, Isaac, Sir, 1642–1727 I. Title

BL241. B320 2015

Manufactured in the U.S.A.　　　　　　　　　　　　　　　　　　　　　　10/02/2015

I dedicate this book to the memory of my parents Russell and Annie Brennan, who passed away during completion of this book. Your wisdom, prayers and encouragement are always missed.

Contents

Acknowledgments | xi

Introduction | xiii

CHAPTER 1   Divine Agency: A Source of Unresolved Issues between Theology and Science | 1

False Starts at Conciliatory Dialogue

Complexity of the Theology and Science Dialogue

Can the Reality of God's Personal Interaction with Humans Be Maintained?

Technical Issues Related to the Theology and Science Dialogue

Divine Agency Develops from Three Factors Commonly Understood in Early Modernity and the Possibility of an Alternative

CHAPTER 2   Divine Agency, Inspiration, Perfection, and Generic Theology | 23

The Convergence of Augustinian Inspiration and Perfect-Being Theology in Early Modern Science's Synthesis of an Impersonal Description of Divine Agency in the World.

Perfect-being Theology Remains a Contemporary Issue

Perfect-being Theology at the Beginnings of the Modern Period until the Mid-nineteenth Century

CONTENTS

The Two Books of God's Revelation as a Factor in Early Modern Understanding of Divine Agency

Inspiration as Guarantee of God's Action—the Third Factor

Augustine's Description of Inspiration by Way of Tertullian

Incarnational Divine Agency and Inspiration

CHAPTER 3 Newton and God/Providence Inspiring the Universe | 92

Newton as Theologian

Newton Studies: Open-ended and Controversial

Aether and Spirit

Matter in a Nutshell: Permeable to the Spirit

Cosmic Strings (after a Fashion): Newton's Gravity

The Sensorium of God

(Newton) Clarke-Leibniz Correspondence

God's Law Revealed by Augustinian Ekstasis

Inspiration of Infinite Space

After Newton

Vestiges of Divine Perfection in Nature

CHAPTER 4 Divine Agency Implying Perfection and the Soul | 147

Perfection as Precondition Challenged by Science

Darwin: Perfection No More

Huxley: Metaphysics No More

The Legacy—the Shape of the Stumbling Block to the Dialogue between Theology and Science

CHAPTER 5 Describing Divine Agency in Humans Pneumatologically and Christologically Beginning with Christ | 195

Barth's Non-Augustinian Pneumatology

Scholarly Debate on Barth's Pneumatology in Church Dogmatics

Natural Theology and Non-Augustinian Pneumatology

Holy Spirit and Humanity

Barth's Doctrine of Scripture

Barth's Anthropology and Holy Spirit

Why Barth Stops—Mystery and Holy Spirit

Barth, Incarnational Divine Agency, and Resolving One Area of Tension

CHAPTER 6   Dialogue with One Obstacle Removed | 257

Revised Divine Agency and the Dialogue with Science

Revised Divine Agency and Doctrine's Function

Implications for the Current Debate: Foundations of Shifting Sand: Which Assumptions? Whose Analysis?

Bibliography | 271

Acknowledgments

It is difficult to summarize in a few words the influences as you write over a number of years. When working closely with the writings of great scholars of the past, it is important to give them great respect especially when you come to disagree with them. Tertullian and Augustine have influenced the development of Western culture, theology, and even my own faith more than I had realized. Isaac Newton helped to instil a love of learning about the natural world that led to me studying physics and a desire to understand how he saw the world. Karl Barth's thought has and continues to challenge and stretch the faith and thinking of a pietist. I am sure Barth's analysis of the academic debates of the last three centuries has yet to be as evaluated and understood as it deserves. As I write these words, I sit less than a hundred meters from the Arafura Sea on the Gove peninsula on Australia's far north coast conscious that it is about 170 years since the crew of the *Beagle* first mapped these Australian shores after returning Darwin to England. In the two hundred years since Darwin's birth, much is owed to Darwin and Huxley for the wealth of honest and intellectually rigorous writing they have left, which enables the scholar to gain insight into their thought.

My thanks also go to those I have met who have helped me along the way. Special appreciation is reserved for the late Rev. Dr. Colin Gunton who thought a particular idea had merit and was worth pursuing. Thanks to Rev. Dr. Gordon Watson for being a listening ear while developing this topic. Thanks to Rev. Drs. Geoff Thomson and David Rankin for their assistance throughout this project, teaching me to write and keeping me from being side-tracked. To my valuable proof-readers, Dr. John Barrie and Mrs. Valda Edwards.

I wish to thank my wife Louise for her support and encouragement over this last decade, at times coping with three university students in the family at once. In addition, I wish to thank God who continues to inspire and who I am sure never gives up on truth and clarity of thought.

Introduction

ON A NUMBER OF occasions during the last century, the dialogue between theology and science stalled because of unresolved underlying issues between the two disciplines. It is not a case of the now largely discredited notion of science and theology being locked in ongoing conflict. Rather it is in reality that their relationship is and always has been much more complex and intimate. Nonetheless, there remain significant issues where there is substantial, if not disagreement then a failure to reach common ground. It is a concern that in some important discussions one seems to be talking apples and the other oranges. The difficulty is compounded when we realize that for quite a number that discussion is internal as believers seek to reconcile in their own minds and faith their understanding and working commitments in both areas. Despair at the possibility of resolving such underlying issues increases pressure to either abandon the notion of divine action in the world or alternatively to heavily revise the Christian faith. No such revision has received broadly based support across the disciplines.

This book seeks to address one issue—how to describe the hand of God. How can we say how God actually and personally acts in the heart and life of humans and possibly in the world? That is, what is the nature of divine agency? An understanding of divine agency developed out of the interaction of three factors in early modernity. Two factors are already well established as influences, late medieval perfect-being theology and the early modern application of the notion of the two books of God's revelation to the understanding of the natural order. The case is made that the third is the early modern appropriation of the doctrine of inspiration, which contains a description of divine agency in humans, which became applied more generally to divine agency in nature. The description of divine agency that developed presumed the existence of the soul and that attributes of a divine perfect-being must be reflected in the natural order. Both of these assumptions, while generally accepted in the seventeenth century, faced serious challenges by the nineteenth.

The status of this description of divine agency changed from that of unquestioned acceptance among natural philosophers of the seventeenth and eighteenth centuries to becoming a stumbling block to the scientists of the nineteenth century and beyond. If, however, it is possible to describe divine agency, including inspiration, without implying or requiring that perfection or the metaphysical soul is essential, then the underlying issue can be resolved. Therefore, an alternative description of divine agency based in the christological notions of *anhypostasia* and *enhypostasia* is proposed to overcome these problems. This proposal warrants serious consideration only if it is theologically coherent and remains plausible while resolving or avoiding a range of known difficulties. The last section of the book establishes this coherence and plausibility.

The proposal sets out to change the relationship between the three factors: inspiration, divine perfection, and the notion of the two books of God's revelation. In early modernity each of these three factors could be expressed generically, without reference to Christology or the Trinity. It is argued that this is problematic. Augustine's description of inspiration and its understanding of divine agency in humans could be and was re-expressed generically without referring to who God is, rather as a good all powerful divine being. Augustine's description can be traced back through Tertullian and be shown to draw on Aristotelian and classical medical ideas including those of the philosopher Cleanthes and the gynaecologist Soranus of Ephesus. The Augustinian description of inspiration understands the soul to be a metaphysical element of a human that necessarily is stood aside during the direct action of the Holy Spirit. Therefore, *ekstasis* is automatic when the Holy Spirit acts. Thus understood inspired divine agency is solely God's action and thus perfect, as epitomized in the production of Scripture.

Newton extended such a non-Trinitarian revision of the Augustinian notion of divine agency in humans to divine agency in the world. In spite of being cautious about publishing it, Newton firmly believed the mind of God relates to the universe as a sensorium in a manner similar to the Aristotelian understanding of the way the human mind relates to the sensorium of the body—its five senses, as well as the abilities to use memory and to command movement. This was Newton's way to ensure that God as the Lord of all times and places. Moreover, Newton's analogy becomes complete only if the mind of God by the Holy Spirit stands aside a fictive mind of the universe in the same way that the Holy Spirit was understood to stand aside the human mind in Augustinian *ekstasis* inspiration.

Newton's scientific successes led to a growing confidence in human ability to understand the laws of God in nature, thereby spurring an interest in natural philosophy. This same success later became an obstacle to interaction between theology and natural philosophy or "science" as it later became known during the nineteenth century. Through the better part of three centuries the unquestioned essential foundations of the Christian faith were thought to include divine perfection, which would be expressed in an harmonious nature in which all creation had purpose and which was the best of all possible worlds. Paley, for example, asserted that the perfect harmonious design in nature was a proof of the Christian faith. A young devotee of Paley's was to later turn him on his head. This was Charles Darwin. While personal tragedy led to Darwin's rejection of traditional Christianity, he rationalized this as a rejection of Paley's argument that perfect harmony proves God's purpose. Darwin's supporter Huxley further rejected traditional Christianity as he described evidence that refuted Aristotelian metaphysical anatomy.

Given that assumed perfection and metaphysics do not hold, this led to serious questioning of the reality of divine agency. However, divine agency is meant to describe the nature of God's personal contact with humans. As such, theologians cannot easily abandon the notion. Nevertheless, it is possible to highlight a formal logical fallacy at work at the root of this "impasse" between theology and science. The impasse can only hold if it is true that divine agency in the world must arise from the description of divine agency in the Augustinian manner and a consideration of divine perfection. My argument is that revision of inspiration and divine agency are needed, not their abandonment. The theological task becomes to offer an account of divine agency not linked to presumptions regarding perfection or metaphysical anatomy.

In the proposed incarnational description of divine agency, the Holy Spirit's action in humans derives from the unique action of the Holy Spirit in the humanity of Christ.

Bringing the incarnational description of inspiration into conversation with the pneumatology of Karl Bath determines whether the proposed incarnational description warrants serious consideration. Barth's incarnational or christological Pneumatology offers an account of the Holy Spirit's work, which also is not dependant on perfection or metaphysical assumptions. Engaging with Barth establishes that the description of divine agency proposed is worthy of serious consideration. The incarnational description also cautiously goes beyond Barth in providing useful detail for resolving issues underlying the development of the dialogue between theology and science. This revised description ceases to depend on any given theological understanding of creation and anthropology. As a description of the Holy Spirit's work, it can be seen as an element of a consistent broader pneumatology rather than as a special case to be treated in isolation.

The revised proposal offers the possibility of resolving one significant underlying issue thus renewing the theology/science dialogue by removing a stumbling block. While insufficient in itself, this revision is a necessary step in providing such a renewed basis for dialogue between theology and science.

CHAPTER I

Divine Agency

A Source of Unresolved Issues between Theology and Science

THE RELATIONSHIP BETWEEN THEOLOGY and science is complex and intertwined. The relationship is not simply one that can be described one-dimensionally as conflict or equal and separate. At times historically they have been close to the point where at one has almost depended on the other. Significant issues such as the question of how best to understand and describe divine agency is interconnected with important assumptions and traditions of thought in science as well as in theology. Further there are additional connections among those assumptions and traditions that deserve careful thought and examination so that a clear and sound description of divine agency can be developed that can make sense to both theology and science. Ultimately, if God exists and is personal as Christians believe, then God must act both in the world and in humans. If there is to be a lasting conciliatory dialogue between theology and science, then establishing how it is that God acts is one particular question needing an adequate answer.

False Starts at Conciliatory Dialogue

Peter Bowler has identified three attempts at conciliatory dialogue between theology and science during the twentieth century. The first was in the century's early decades, which he discusses at length, the second was from

1945 to the 1960s with the third and present beginning in the century's last decade.[1] The first two attempts at conciliatory dialogue stalled because of unresolved underlying issues between the two disciplines. Bowler pessimistically observes that these "seem to reflect the fluctuating balance of power between secularizing and traditional forces with our society, and if this is so, we can surely learn something of value from the debates of earlier decades—if only the futility of expecting the underlying issues ever to be resolved."[2] Bowler is not optimistic that even in the present dialogue that resolution can be reached. Despair about the possibility of resolving such underlying issues increases pressure to abandon conciliatory attempts at finding a true meeting of the minds in dialogue. The temptation has been to seek revision, usually of the Christian faith rather than of science, or to relegate such dialogue to a place of little importance. Unsurprisingly perhaps, no such revision of science or theology has received broadly based support across the disciplines.

The temptation to minimize the importance of such dialogue is problematic, as Nicholas Lash has suggested: "Few of us would survive for long if we seriously supposed our deepest convictions to be illusory or false."[3] Lash proposes that the serious engagement with the dialogue between theology and science is a matter of truthfulness "integral not only to morality but to sanity."[4] For a Christian seriously engaging with the claims of science at "the practical level, this is a question about likely or appropriate forms of survival (if any) of religious belief and practice."[5] The process is not necessarily straightforward, as the sociologist Eileen Barker has pointed out: "[B]oth personalised and institutionalised theologies can encompass the most extraordinary ragbag of facts, opinions and beliefs which happily coexist, apparently quite oblivious of what to others are the most glaring inconsistencies."[6] On the other hand, "as we begin to contemplate the popular image of modern science a further bewildering assortment of contradictions, mysteries and paradoxes emerges. Science seeks out the immutable laws of the universe yet reveals the universal principles of indeterminacy and uncertainty."[7] A meeting of the minds in conciliatory dialogue between theology and science

---

1. Bowler, *Reconciling Science and Religion*, 4.
2. Ibid. 5.
3. Lash, "Theory Theology and Ideology," 209.
4. Ibid.
5. Ibid.
6. Barker, "Science as Theology—the Theological Functioning of Western Science," 263.
7. Ibid.

often naively presumes that serious dialogue has already occurred within the mind of the believer. It is presumed they have sought to take seriously the claims of their faith as well of those of the science they have come to know. Barker's summary of religious belief held by scientists suggests that it would be wise not to presume too much on such internal dialogue. Rather, Barker highlights that the variation and spread of belief is as broad in this group as it is among the wider community. Scientists engaging in theological debate hold opinions varying from fundamentalism to atheism.[8]

Irrespective of the complexities of the theology and science dialogue that will be outlined later, is it possible to resolve the underlying issues that continue to keep in contradiction, mystery, and paradox an individual's ragbag collection of facts, opinions, and beliefs? Is such resolution even possible? While this cannot be answered fully, this book suggests that one important underlying issue can be resolved: alleviating part of the pessimism to which Bowler refers. At the heart of the issue is the question of divine agency, how to describe God's direct action in the world and in humans. It is widely assumed, even when disputed, that such divine agency will somehow reflect God's perfect-being.

It will be argued that this assumption arose historically from an understanding of divine agency in the world developed with a conjunction of three factors in early modernity. The first factors date from the late medieval period and are relatively well documented. These are, firstly, how the divine perfections were understood in relation to nature and, secondly, how the notion of the two books of God's revelation (Scripture and nature) affects understanding of the natural world. This book argues for serious consideration of a third: the doctrine of inspiration and its encapsulated description of divine agency in humans. This description, it will be argued, came to be used more broadly to describe divine agency in nature. During this historical development in early modernity only one description of inspiration was in use—Augustine's *ekstasis* description.

This particular understanding of divine action assumes a particular arrangement to human anatomy. Firstly, that humanity has a spiritual component or soul and secondly that when God acts in a person this soul is stood to one side. The implication of the soul being stood aside in that the ensuing action is understood to be God's own action through the human. In "pure" inspiration it is God's rather than the human will at work. In as much as such action is purely God's own, such divine agency is understood to be perfect. This has two logical implications which color further discussion: firstly, that there exists a metaphysical component to human anatomy, the

---

8. Ibid., 267–68.

soul, which becomes an essential element to anthropology; and secondly, that divine action in this manner overcomes the limitations of human finitude and is able to achieve perfection. These initially offered a useful way to describe how divine action could be perfectly reflected in an imperfect world. They act to complement early modern understanding of the divine perfections and the notion of the two books. Such a mechanism of divine agency provided a useful tool for describing the supposed ideal nature of God's actions in creation as natural philosophy developed. The supposition that nature, if read using the correct methods, could reveal more of the purposes and grandeur of God spurred to the disciplined and thorough study of nature.[9] In particular Augustinian anatomy of the soul and *ekstasis* became demonstrably foundational for Newton's understanding of matter.

However, these suppositions, contribute to what Buckley has described in the nineteenth century as "the tensions and contradictions within the various forms of natural theology."[10] Disciplined and thorough study of nature raised issues with the notion of ideal perfection of divine action in nature. This led to revision of the notion of perfect divine action with it coming to be discussed in terms of teleological perfection.[11] That is, that God's ends would be perfectly met rather than there being perfection in every detail. However, even the possible grounds for such supposed teleological perfection become further eroded as the sciences developed. It will be shown that Darwin particularly rejected all notion of teleological perfection. His advocate Huxley further raised serious questions about the existence of a metaphysical soul, leaving both assumptions in dispute.

If a Christian description of divine agency must continue to rely on either assumption, perfection in divine agency or the existence of a metaphysical soul, then the description risks becoming problematic. Divine agency would continue to be an unresolved issue in the dialogue between theology and science.

Nonetheless, in addition to the evolution of their use historically, these two assumptions also allowed the discussion of God's actions in nature, including humans, to be discussed generically, independently of any reference to who God may be. Buckley notes surprise at the minimal response to Newton's protégé Clarke's defence of true religion without mention of,

---

9. Harrison, *The Bible, Protestantism and the Rise of Natural Science*; Brooke, "Reading the Book of Nature"; Hess, "God's Two Books of Revelation: The Life Cycle of a Theological Metaphor."

10. Buckley, *At the Origins of Modern Atheism*, 358.

11. Clayton, *The Problem of God in Modern Thought*; Passmore, *The Perfectibility of Man*.

"Christology or religious experience."[12] More surprising in the contentious air of that debate is that no one confronts his omission and that this absence "stirs nary a tremor."[13] That divine action could be discussed generically, was demonstrably attractive in an era hostile to any real or perceived departure from orthodoxy. It was particularly attractive to a significant number of leading natural philosophers like Clarke who had adopted heterodox Christologies. Such generic description of divine agency as Divine Providence or simply Providence also helped to avoid controversies around the doctrine of the trinity. By the early nineteenth century this kind of generic description had become widely used. A good example is that the deist William Paley's texts became the set texts at Cambridge for the first major examination.[14]

It will be shown that these theological notions are ultimately self-contradictory. Whilst they contributed to the development of science, both assumptions also contain within themselves the seeds of their mutual destruction. It will be argued that divine agency relying on these assumptions is not the only viable description. If it is possible, as will be argued, to describe divine action, including inspiration, without implying that perfection or metaphysics is essential, then this particular underlying issue can be resolved.

Drawing on the eastern theological tradition, an alternative description is posed developing a description of God's intimate and personal communication by the Holy Spirit from the christological notions of *anhypostasia* and *enhypostasia*. In this the intimate and personal contact of God's Holy Spirit with the human spirit is shaped to human need and limitations in the humanity of Christ. This revised incarnational description makes no assumptions about perfection or metaphysics.

The proposed incarnational description will undergo examination in conversation with the Pneumatology of Karl Barth to order to establish whether it is worth serious consideration. Barth's incarnational or christological Pneumatology also offers an account of the Holy Spirit's work that is not wedded to perfection or metaphysical assumptions. The extensive nature of Barth's theology also allows the study of divine agency in relation to broader academic concerns. These concerns include those underlying the development of the dialogue between theology and science: theological understanding of creation and anthropology; consistency in Pneumatology; and the doctrine of Scripture.

---

12. Buckley, "Science as Theology," in *At the Origins of Modern Atheism*, 354.
13. Ibid.
14. Paley, *A View of the Evidences of Christianity*; Desmond and Moore, *Darwin*, 64.

This proposal warrants serious consideration if it is theologically coherent and remains plausible while resolving or avoiding a range of known difficulties. The last section of this book establishes this coherence and plausibility. As a revised description of divine agency, this proposal will be shown to avoid past problems, while being able to stand in engagement with a breadth of theological issues. Such a revision is one necessary step of many needed to resolve underlying issues in the breadth of the dialogue. Doubtless, divine agency is not the only reason for underlying unresolved issues between theology and science. My aim is to address divine agency as one of the unresolved issues needed to enable conciliatory debate between theology and science to continue. The proposal that the incarnation is constitutive for divine agency in humans will be shown to substantially alter the mix of these ideas and not to lead to the same kinds of tensions and contradiction to which Buckley refers.[15] This, however, leaves open the question of divine agency in the world though it is possible that such a description might suggest how the incarnation might be considered constitutive of divine agency in general.

It needs to be stated clearly from the outset that the relationship between theology and science is complex and is not well served by one-dimensional descriptions. To say they are in conflict is simply not supported by the literature. There are harmonies. There are areas of independence. There are areas of constructive interaction. These are demonstrable between the disciplines and even exist within each discipline and in dialogues on specific issues. Also disputes and methodological issues exist within each discipline, which are not connected with the dialogue.

Complexity of the Theology and Science Dialogue

As Bowler has indicated academic debate between theology and science has grown afresh since the last decade of the twentieth century. While this may seem to give theology an apparently stronger voice, on closer inspection this may not always be the case. Peterson argues that theology should be considered an equal partner in the theology science dialogue, but notes as a difficulty theology's absence as a discipline from the contemporary university.[16] Nonetheless, even if such equality is in question, a brief literature review indicates there is a great deal of cooperation, goodwill and attempts at understanding between theology and the sciences. The outcome of the interactions varies considerably. There are examples of mutual understanding

---

15. Buckley, *At the Origins of Modern Atheism*, 358.
16. Peterson, "In Praise of Folly? Theology and the University."

and even necessary interdependence; however, there remain some issues where differences between theology and science remain apparently intractable. In spite of significant concord being reached, these intractable issues still hold the potential to derail dialogue yet again.[17] This book suggests a way to resolve one such intractable case. In doing so it may offer hope that it may be possible to advance the dialogue beyond such points of intellectual stalemate or conflict.

While many scholars have commented on the shape the dialogue takes or should take,[18] Barbour has been influential in offering four ways of understanding the interaction between science and religion: conflict, independence, dialogue and integration.[19] Though criticized as limiting and misleading,[20] Barbour's categories are often used in sociological analyses of as the nature of the debate and the spread of academic thinking and belief.[21] Brown particularly shows among scientists that the range of contemporary beliefs remains similar to the range held in 1910. As Brown's results have been confirmed[22] it has been suggested that little has changed in the debate during the last century, whether it be due to lack of critical self-analysis, failure to resolve key issues, or failure of solutions to gain wide support or interest. Worthing has indicated that many scientists still enter the dialogue with theologically conservative notions like those of the nineteenth century, which directly affect their expectations of the shape of divine agency in the world.[23]

In spite of Barbour and others championing conciliatory models for the shape of the dialogue, the public extremes endure.[24] Irrespective of their

---

17. Bowler, *Reconciling Science*, 1–20, 411–20.

18. Chung, "Karl Barth and God in Creation: Towards an Interfaith Dialogue with Science and Religion"; Marcum, "Exploring the Rational Boundaries between the Natural Sciences and Christian Theology"; Moritz, "Science and Religion: A Fundamental Face-Off, or Is There a *Tertium Quid*?"; Murphy, "On the Role of Philosophy in Theology-Science Dialogue"; Rae, Regan, and Stenhouse, eds., *Science and Theology: Questions at the Interface*; Trenn, "Science, Faith and Design,"; Nebelsick, *Theology and Science in Mutual Modification*; Peacocke, *The Sciences and Theology in the Twentieth Century*.

19. Barbour, *When Science Meets Religion*.

20. Brooke, "The Changing Relations between Science and Religion"; Cantor and Kenny, "Barbour's Fourfold Way: Problems with His Taxonomy of Science-Religion Relationships," 1–20.

21. Brown, "The Conflict between Religion and Science in Light of the Patterns of Religious Belief among Scientists."

22. Case-Winter, "The Question of God in an Age of Science "; Larson and Witham, "Scientists Are Still Keeping the Faith."

23. Worthing, *God, Creation and Contemporary Physics*, 29–30, 159–68.

24. Christian creationism and materialistic atheism view each other as the "root of

relative academic merits there continues vociferous public and academic debate. In North America this has been part of what has been termed, "culture wars." Studying the history of the debate, the persistence of forms of academic creationism as well as their ongoing developments in social and political thought led Numbers to revise and greatly expand his study on these schools of thought arguing for their serious academic consideration without necessarily agreeing with their arguments. Conciliatory examples of dialogue do exist[25] and involvement of theology is indispensable in relation to ethical scientific research.[26]

In terms of Barbour's four categories of interaction—conflict, independence, dialogue and integration—all are present, though it would be fair to say that discussions between theology and cosmology show more dialogue and integration[27] whereas those between theology and the life sciences show more conflict and independence.[28] While there are instances where common ground can be identified such as natural selection favouring the development of ethical or theological notions such as altruism,[29] there

---

all evil." Dawkins, *The God Delusion*; Shanks and Dawkins, *God, the Devil, and Darwin: A Critique of Intelligent Design Theory*. 10. Lambert, "Fuller's Folly, Kuhnian Paradigms, and Intelligent Design"; Smedes, "Social and Ideological Roots of 'Science and Religion': A Social-Historical Exploration of a Recent Phenomenon." Numbers, *The Creationists: From Scientific Creationism to Intelligent Design*. Original and revised editions. The revision greatly expanded the book from 436 to 606 pages.

25. Pannenberg, *Toward a Theology of Nature*; Murphy, "What Has Theology to Learn from Scientific Methodology?"; Murphy, "Science as Goad and Guide for Theology,"; Ruse, "An Evolutionist Thinks About Religion,"; Edwards, "Christology in the Meeting between Science and Religion: A Tribute to Ian Barbour"; Jackelén, "What Theology Can Do for Science"; Burtt, *The Metaphysical Foundations of Modern Physical Science*.

26. Klinefelter, "E. O. Wilson and the Limits of Ethical Naturalism"; Kuczewski, "Two Models of Ethical Consensus, or What Good Is a Bunch of Bioethicists"; Shults, "Anglo-American Postmodernity: Philosophical Perspectives on Science, Religion, and Ethics."

27. Ross, *The Creator and the Cosmos*; Craig and Smith, *Theism, Atheism and Big Bang Cosmology*; Jastrow, *God and the Astronomers*; Stoeger, "Cosmology and a Theology of Creation"; Stoeger, "Science the Laws of Nature and Divine Action."

28. Wilson, *Consilience*; Young, "Can the Creationist Controversy Be Resolved?"; Ruse, "John Paul II and Evolution"; Edwards, "Evolution and the Christian God"; Barbour, "Evolution and Process Thought"; Haught, "In Praise of Imperfection"; Hewlett and Peters, "Why Darwin's Theory of Evolution Deserves Theological Support"; Lennox, *God's Undertaker: Has Science Buried God?*; Ashton, *In Six Days*.

29. Dawkins, *The Selfish Gene*, viii. Dawkins would himself reject that there was any theological implication of this effect.

remain apparent impasses or contradictions such as the theological notion of purposeful eternity versus the heat death of the universe.[30]

Hints of the existence of commonality in dialogue have fuelled interest in the revision of theology and less commonly science to either overcome particular impasses or improve that which is held in common.[31] One commonly used type revises theology using Whitehead's process theology.[32] Bowler notes similar Whiteheadian influence in the work of Waddington, Eddington, Barnes, Fisher, Needham, Morgan, Morrison, Inge, Thornton and Temple. It is salient to note Bowler's warnings arising from these and other attempts to harmonize theology and science in the early twentieth century. These he argues were prone to two related errors. "A relatively small number of influential writers were able to present an interpretation of science that was almost certainly out of touch with what the majority of working scientists thought."[33] Secondly, that theological revisions linked to a particular theological school or theory failed to win support as that school or theory lost or failed to achieve prominence.[34] Any revision hoping to succeed must win wide acceptance and be relevant among both theologians and scientists.

It is generally accepted that theological understandings influenced the historical development of science from its roots in seventeenth-century natural philosophy to the nineteenth century.[35] A typical summary is that of Hess who notes,

---

30. Polkinghorne, *Science and Christian Belief*, 162–70.

31. Bowler, "Development and Adaptation: Evolutionary Concepts in British Morphology"; Bowler, "Evolution and the Eucharist: Bishop E. W. Barnes on Science and Religion in the 1920s and 1930s"; Bowler, *Reconciling Science and Religion: The Debate in Early Twentieth Century Britain*; Whitehead, *Process and Reality: An Essay in Cosmology*; Worthing, "God, Process and Cosmos: Is God Just Going Along for the Ride?"; Peacocke, "Science and the Future of Theology: Critical Issues."

32. Barbour, *When Science Meets Religion*; Barbour, "Evolution and Process Thought"; Needham, *Science, Religion, and Socialism*; Needham, *Science Religion and Reality*; Bowler, *Reconciling Science and Religion: The Debate in Early Twentieth Century Britain*, 80, 104, 154, 171–72, 241, 275, 280, 304, 306.

33. Ibid., 420.

34. Ibid. 411–18.

35. Lindberg, *The Beginnings of Western Science*; Numbers and Lindberg, *God and Nature: Historical Essays on the Encounter between Christianity and Science*; Brooke, *Science and Religion: Some Historical Perspectives*; Brooke, "The Changing Relations between Science and Religion"; Brooke and Cantor, *Reconstructing Nature: The Engagement of Science and Religion*; Brooke, "Reading the Book of Nature"; Henry, *The Scientific Revolution and the Origins of Modern Science*; Harrison, *The Bible, Protestantism and the Rise of Natural Science*; Harrison, "Curiosity, Forbidden Knowledge, and the Reformation of Natural Philosophy in Early Modern England"; Harrison, "The Book of

> Pervading the tradition of natural theology in the Christian West has been the theme of "God's two Books." This metaphorical pairing the "book of nature" and the "book of Scripture" expressed the medieval and early modern conviction that the divine existence and wisdom are clearly revealed by a pair of complementary sources... How did the nineteenth century development of evolutionary biology and historical biblical criticism—both of which so profoundly inform our contemporary dialogue—lead to its abandonment or drastic modification?[36]

Harrison suggests that science developed as similar interpretive rigor to read the book of nature as that used in relation to the book of Scripture. Harrison's thesis is that in the process of adapting rigorous methods for reading the book of nature the assumptions which underpinned interpretation of the book of Scripture were also applied to the book of nature. This current discussion builds on Harrison's argument that that an understanding of divine agency in world developed through the application of related assumptions.

There exist other descriptions of how theological presuppositions influenced the development of modern science. These have been often used but have been challenged and will continue to be criticized here. Two influential theories need particular mention. The first is Merton's 1938 thesis suggesting that Puritanism was necessary to the rise of modern science.[37] Brooke and Harrison detail how Merton fails to encompass the broader protestant influence of actual practitioners of science and Harrison leaves Merton to suggest more specifically that Protestant interpretation of texts was a major catalyst.[38] The second influential thesis is that of Foster who asserted the Calvinistic notion of Divine voluntarism has been a spur to the

---

Nature and Early Modern Science"; Harrison, "Religion, the Royal Society, and the Rise of Science"; Lightman, *The Origins of Agnosticism: Victorian Unbelief and the Limits of Knowledge*; Hooykaas, *Religion and the Rise of Modern Science*.

36. Hess, "God's Two Books of Revelation: The Life Cycle of a Theological Metaphor."

37. Merton, *Science, Technology and Society in the Seventeenth Century England*. Critics include Osler, "Mixing Metaphors: Science and Religion or Natural Philosophy and Theology in Early Modern Europe"; Brooke, *Science and Religion: Some Historical Perspectives*, 110–16; Henry, *The Scientific Revolution*, 93, 94; Harrison, *The Bible, Protestantism and the Rise of Natural Science*, 8; Harrison, *The Bible, Protestantism and the Rise of Natural Science*; Harrison, "The Book of Nature and Early Modern Science"; Harrison, "'Science and Religion': Constructing the Boundaries"; Greaves, "Puritanism and Science: Anatomy of a Controversy."

38. Harrison, "Voluntarism and Early Modern Science"; Harrison, "Was Newton a Voluntarist?"

investigation of the natural order.[39] While this complex theory has significant usage it has received some criticism that it reaches too far.[40] Harrison suggests its dismissal. Among Harrison's reasons is that voluntarism is not able to be demonstrated to have influenced actual key historical figures.[41] While Foster's thesis seems plausible, actual examination of the writings of supposed voluntarists, such as Newton, reveals content which directly contradicts Foster's assertions.[42]

Can the Reality of God's Personal Interaction with Humans Be Maintained?

As this book will focus on the influence of one particular understanding of divine agency—how God acts through the spirit in humans—it is worth noting that another open question both theologically and scientifically is the nature of spirit and God.[43] The nature of spiritual existence and its shape is a topic of ongoing study and conjecture which extends to investigations regarding the nature of the soul or even is existence.[44] The existence of a

39. Foster, "The Christian Doctrine of Creation and the Rise of Modern Science"; Foster, "Christian Theology and Modern Science of Nature (I.)"; Foster, "Christian Theology and Modern Science of Nature (II.)"; Oakley, "Christian Theology and the Newtonian Science: The Rise of the Concept of the Laws of Nature."

40. Davis, "Christianity and Early Modern Science: Beyond War and Peace?"

41. Harrison, "Voluntarism and Early Modern Science"; Harrison, "Was Newton a Voluntarist?"

42. See chapter 3 on Newton.

43. Green, "Restoring the Human Person: New Testament Voices for a Wholistic and Social Anthropology"; Murphy, "Darwin, Social Theory, and the Sociology of Scientific Knowledge"; Murphy, "How Physicalists Avoid Being Reductionists"; Murphy, "Why Christians Should Be Physicalists"; Clayton, "Biology, Directionality, and God: Getting Clear on the Stakes for Religion—Science Discussion"; Clayton, "The Emergence of Spirit: From Complexity to Anthropology to Theology"; Conway, "Defining 'Spirit': An Encounter between Naturalists and Trans-Naturalists"; Work, "Pneumatological Relations and Christian Disunity in Theology-Science Dialogue"; Yong, "Discerning the Spirit(s) in the Natural World: Toward a Typology of 'Spirit'"; Polkinghorne, "Physics and Metaphysics in a Trinitarian Perspective"; Simmons, "Quantum Perichoresis: Quantum Field Theory and the Trinity"; Yong, "The Spirit at Work in the World: A Pentecostal-Charismatic Perspective on the Divine Action Project"; Pannenberg, "God as Spirit—and Natural Science."

44. Green, "Restoring the Human Person: New Testament Voices for a Wholistic and Social Anthropology"; Russell et al., *Neuroscience and the Person: Scientific Perspectives on Divine Action*; Masters and Churchland, "Neuroscience and Human Nature the Engine of Reason, the Seat of the Soul: A Philosophical Journey into the Brain"; Spezio, "Interiority and Purpose: Emerging Points of Contact for Theology and the Neurosciences"; Watts, *Science and Theology*; Barrett, "Is the Spell Really Broken?

human metaphysical spiritual element or soul has been an apparent mainstay of the Christian faith. Whilst Green, Murphy and others argue that this need not be the case, the answers to many theological questions presuppose the existence of a metaphysical spirit or soul. If there is no metaphysical soul then many theological descriptions will need revision or even abandonment. This is critical for this discussion which focuses on one aspect of God's work through the spirit, namely the agency by which God communicates knowledge of God by the action of the Holy Spirit to or through the human spirit. The manner of God's self-communication to humans has traditionally been described as involving a metaphysical human soul or spirit. If such divine communication depends on there being a metaphysical soul, then the Christian faith stands or falls on the health of that premise. Worryingly for this premise, neurobiological studies have located many attributes previously considered spiritual and hence metaphysical within the biochemistry of the brain. Such a rational and strong challenge to the existence of metaphysical soul is demonstrably not new.[45] A number of questions might be posed. It could be asked whether there may be a way to describe the soul which answers the challenge of neurobiology and anatomy. Rather than pose a "soul-of-the-gaps" this book will ask whether the nature of divine self-communication to humans can be described in a manner which operates independently of any given metaphysical anthropological theory. What is at stake in this question is whether divine communication to humans can actually occur as intimately and personally as Christian theology has contended. If such communication is predicated on God's contact with a metaphysical soul and there proves to be no such entity, then God can only be known by indirect means and traditional Christianity becomes problematic.

The question to be addressed is whether it is possible to describe the agency God's intimate and personal communication by the Holy Spirit independently of metaphysical anthropology. Drawing on the eastern theological tradition, an alternative description is posed developing a description of God's intimate and personal communication by the Holy Spirit from the christological notions of *anhypostasia* and *enhypostasia*. In this the intimate

---

Bio-Psychological Explanations of Religion and Theistic Belief"; Dodds, "Hylomorphism and Human Wholeness: Perspectives on the Mind-Brain Problem"; Jeffreys, "The Soul Is Alive and Well: Nonreductive Physicalism and Emergent Mental Properties"; Jeffreys, "A Counter-Response to Nancey Murphy on Non-Reductive Physicalism"; Murphy, "Response to Derek Jeffreys."

45. Brennan, "Has the frog human a soul?"; see chapter 4.3 on Huxley. Huxley, "On the Present State of Knowledge as the Structure and Functions of Nerve"; Huxley, "Has a Frog a Soul?"; Huxley, "On Sensation and the Unity of Structure of Sensiferous Organs."

and personal contact of God's Holy Spirit with the human spirit is shaped to human need and limitations in the humanity of Christ.

It will be argued that the appropriate theological context for discussing the agency of divine communications with human beings is in relation to the question of inspiration. This is a rather more general usage of inspiration than in relation to the doctrine of Scripture which has been inspiration's main focus in three of the last four centuries. A new place will be suggested for inspiration within Pneumatology in general apart from being in relation to the doctrine of Scripture. Why this particular aspect of the Holy Spirit's work should be considered more broadly will be proposed by reference to the development of the terminology of inspiration in the early church.

Technical Issues Related to the Theology and Science Dialogue

Before describing how this book will develop it is appropriate to highlight some technical issues which affect study of these disciplines. Failure to recognize these issues has led research to incorrect results, to overlook important historical detail and relationships as well as led to inappropriate generalized assertion of conclusions. The two areas of particular interest are historiographical bias in the histories of science and the logical fallacy of affirming the consequent. The third technical issue involves the place of inspiration as a doctrine within the broader question of the place of doctrine within theology.

Bias and History of Science

Imre Lakatos adapting Kant commented "Philosophy of science without history of science is empty; history of science without Philosophy of science is blind."[46] This comment is particularly relevant as much history of science has been blind to well-known biases which have adversely affected the study of some important figures. A case in point is the study of Isaac Newton which has been clouded by multiple revisionist histories and serious ongoing politicized debate.[47]

---

46. Lakatos, "History of Science and Its Rational Reconstructions," 102.

47. Fara, *Newton: The Making of Genius*; Jacob, "Introduction"; Noakes, "Recreating Newton: Newton Biography and the Making of Nineteenth Century Science"; Osler, "The New Newtonian Scholarship and the Fate of the Scientific Revolution."

Anachronism is often overlooked. Applying a term or an idea to a time in which it is not used is one error that should be obvious but is often missed. The use of the terms science and religion serve to illustrate the point as they are often used of debates centuries into the past. "Science" in its modern usage was first applied to the discipline in the mid-nineteenth century by William Whewell, similarly "religion" before this period referred to personal faith rather than a system of belief.[48]

Regarding terminology, it is important to note that in Newton's period, the seventeenth and eighteenth centuries, he and his colleagues are correctly termed Natural Philosophers. It is only by the mid nineteenth century that the newly coined term "Scientist" is applied. The term was coined by William Whewell. There is a transition from the seventeenth-century polymath who might, like Galileo, be expected to be expert in mathematics, astronomy, astrology, alchemy and music to the specialist like Darwin who devoted years to a much narrower discipline such as barnacles within the new science of biology. Ironically, it was this commitment to a detailed methodological and exhaustive study of the book of nature which allowed the detailed study of biology to have become part of science by the nineteenth century. Mere animal husbandry was deemed to be beneath the interest of the natural philosopher in the seventeenth. Hence the notion of the two books ironically prompted the detailed study of nature, a study which will be shown to later sow the seeds of its own demise.

In addition to anachronism there are other well documented historiographical biases which particularly affect histories of science; Whigg histories of onward ever upward progress into the shining present fails to appreciate the past on its own terms;[49] presentism, a specific type of anachronism, where concerns, motivations, terms and ideologies in the past are not interpreted in relation to their past use but in relation to present theories or ideologies, e.g. Merton interpreting Puritan thought by "obviously" superior 1938 science[50] and; the myth of the heroic rational and moral scientist working in ideal solitude to further knowledge.[51] These types of biases

---

48. Harrison, "'Science and Religion': The Constructing the Boundaries."

49. McEvoy, "Positivism, Whiggism, and the Chemical Revolution: A Study in the Historiography of Chemistry."

50. Kragh, *An Introduction to the Historiography of Science*, 47. Merton, *Science, Technology and Society in the Seventeenth Century England*. See also Osler, "Mixing Metaphors: Science and Religion or Natural Philosophy and Theology in Early Modern Europe," 96–99.

51. Appleby, Hunt, and Jacob, *Telling the Truth about History*, esp. 15–51; Jeans, *The Growth of the Physical Sciences*; Lodge, *Pioneers of Science*; Yeo, *Defining Science: William Whewell, Natural Knowledge, and Public Debate in Early Victorian*.

lead to the rejection of historical data that do not fit the explanatory theories and can lead to the imposition of ideas resulting in seemingly satisfying contemporary theories which have little to do with historical fact. Draper and White's largely discredited warfare myth[52] fails in part by ignoring these kinds of bias.

Any discussion dealing with the history of the interactions between theology and science and their antecedents needs to remain conscious of such sources of bias. More weight should be given to primary sources than later theory.

A Logical Fallacy—Affirming the Consequent

Logical fallacies obviously lead to problematic reasoning and incorrect conclusions. A key logical fallacy related to the dialogue is affirming the consequent.[53] This logical fallacy can occur in Whigg histories of science as such histories tend to omit details of history that do not fit the orderly progression and improvement of ideas. This fallacy is characterized by concluding that a consequent outcome must be the result of a particular chain of events. In a simple form this would be,

> e.g.: A) *If a car runs out of fuel it stops.*
>
> B) *Your car has stopped.*
>
> C) *The false conclusion—your car must have run out of fuel.*

This can only be true if lack of fuel is the only possible reason for the car stopping. It is a fallacy because while each logical step may lead to the conclusion the outcomes may well be caused by other means. The Biologist E O Wilson has stated that evolutionary biologists are particularly prone to committing this fallacy.[54] The fallacy lies in concluding that if the answer obtained looks like it is right, then all the steps to get there are right too. It is like saying you took all the right directions to get to your destination no matter how often you got lost or how late you arrive. Conversely, if the conclusion is wrong it is often mistakenly assumed that all steps taken are also wrong.

---

52. Draper, *History of the Conflict between Religion and Science*; White, *A History of the Warfare of Science with Theology in Christendom*.
53. Warburton, *Thinking from A to Z*, 5–7.
54. Wilson, *Consilience*, 94–95.

Affirming the consequent will be an issue twice in the course of this book. The first case involves Foster's theory which purportedly explains the development of Newton's thought. The second case deals with the development of an apparent impasse in which forces a choice between scientific rationality and religious sentiment as the basis for theology in the nineteenth century.

The Place of Doctrine in Theology

The third technical issue is how doctrines work within theology. Divine agency and the doctrine of inspiration do not stand alone from broader questions about the place and functioning of doctrine within theology as an academic discipline and theology's future as an academic discipline. Lash recognises the difficulty of theology's status attributing the challenged academic status of theology to a more general problem. He states that in this last period of modernity westerners are left with the enlightenment legacy of "a crisis of docility." That is:

> Unless we have the courage to work things out for ourselves, to take as true only that which we have personally ascertained or, perhaps, invented then meanings and values, descriptions and instructions, imposed by other people, feeding other people's power, will inhibit and enslave us, bind us into fables and falsehoods from the past. Even God's truth, perhaps especially God's truth, is no exception to this rule. Only slaves and children should be teachable or docile.[55]

This legacy has affected the nature of theological discourse so that

> by the end of the nineteenth-century, Western religious thought found itself trapped by the dominant narrative into an uncomfortable dilemma: either, on the one hand, adopt discredited and outdated particularities of worship, association and belief ("sect," "ghetto" and "dogma" not being labels of approbation); or, on the other, embrace that diffuse religiosity of discourse which suffuses national identity, ambition and public control with a warm glow of transcendent benediction, giving currency (sometimes quite literally!) to the sentiment in God we trust.'[56]

The problem for Christian theology is that if the Christian faith has any basis for making broad public truth claims then such a billabong existence is

---

55. Lash, *Believing Three Ways in One God: A Reading of the Apostles Creed*, 10.
56. Lash, *The Beginning and the End of Religion*, 222.

a denial of the importance of its subject matter. A billabong is an Australian Aboriginal word for an often calm leafy pleasant waterhole which is left behind after a river changes course. In the arid Australian climate, they are prone to suddenly drying out and dying. Because of what it claims to deal with, theology cannot allow itself to remain at the margins of serious academic debate nor be seen merely as an end in itself.

A concurrent difficulty is theological, as terminology used to speak of the work of the Holy Spirit has suffered from both conflation and narrowing. One such example is the almost synonymous use of the term revelation with inspiration or indeed to replace inspiration which has been seen to be a difficult term. One of the difficulties presented by this usage is that a more general enlivening sense of the Holy Spirit's agency becomes confused with the impartation of knowledge or solely with the impartation of propositional truth. Paradoxically, reference to the doctrine of inspiration becomes merely shorthand to describe the narrow horizon of knowledge imparted from the perspective of the inspiration of Scripture. This is far removed from Calvin's usage in which personal inspiration by the Holy Spirit is an act of divine agency which confirms the prior inspiration of Scripture.[57]

There is a question of terminology about how to speak of the work of the Holy Spirit. How should we distinguish the Holy Spirit's action in the human person in general terms as a subset of Pneumatology in general from the Holy Spirit's role in intra-Trinitarian relations or in divine agency in the world or in eschatology? Rowan Williams, commenting that there has been, "a certain poverty in theological reflection on the Holy Spirit in Western Christianity over the last decades,"[58] addresses the personal work of the Holy Spirit with reference to the Johannine concept of Paraclete.

> John sees the Paraclete as active in and with the disciples, moving them towards Father and Son, as well as acting simply *upon* them. The agency of the Paraclete is understood in terms of distance and response rather than simple identification with the agency of Father and Word.[59]

Even in the personal work of the Holy Spirit in humans, this paracletic work, it is possible to distinguish a range of actions. Williams cautions against speaking of this paracletic work too narrowly lest the richness of who God is be lost in the description.

---

57. Calvin, *Institutes* 1.8.7, 1.7.8, 1.13.7, 1.14.7, 1.18.2, 3.20.5, 3.20.42, 4.10.25.
58. Williams, *On Christian Theology*, 107.
59. Ibid., 119.

> The Spirit is associated with the character of Christian existence as such, creating in the human subject response to, and conformation to, the Son. The Spirit's witness is not a pointing to the Son outside the human world, it is precisely the formation of "Son-like" life in the human world; it is the continuing state of sharing in the mutuality of Father and Son; it is forgiven or justified life. . . . The distance between God and the world is transcended . . . And if all this is, in whatever sense, the work of Spirit, it is clear that the association of Spirit exclusively or chiefly with the more dramatic charismata is a misunderstanding.[60]

One may argue that all these actions are interrelated, as indeed is the paracletic work of the Holy Spirit with the Spirit's role and work in general it is useful to differentiate elements such as the Spirit's "inspiring work" or "sanctifying work" or "converting work" or "recreating work." Thus this discussion focuses on the description of the inspiring work of the Holy Spirit and more particularly on the agency of that divine interaction. While presumably such agency would address the production of Scripture, this would only be one aspect of how God by this agency acts to enliven humans to know, to learn and to act in a new manner.

How it is that theology's claims act as doctrines is also a matter of current debate. While Lindbeck notes that most Christian traditions have held that their doctrines are normative and permanent, there has, he claims, developed a contemporary environment of antidoctrinalism in opposition to what developed as a polarisation with theology between treating doctrines as either propositional statements or expressions of subjective pre-cognitive experience.

He argues that a regulative or rule theory for doctrine that restates traditional doctrines "has advantages over other positions"[61] and is essential to enabling theology to continue to have a voice in academic debate and in ensuring the cohesion of the faith itself. "Privatism and subjectivism that accompany the neglect of communal doctrines lead to a weakening of the social groups . . . that are the chief bulwarks against chaos and against totalitarian effort to master chaos."[62] With such revisions to doctrines "it need not be the religion that is primarily reinterpreted as world-views change, but rather the reverse: changing world-views may be reinterpreted by the one and same religion."[63]

60. Ibid., 120.
61. Lindbeck, *The Nature of Doctrine: Religion and Theology in a Postliberal Age*, 73.
62. Ibid., 77–78.
63. Ibid., 82.

Lindbeck is not alone in seeking a revision or renewing of doctrine. Francis Watson, with the aim of rekindling dialogue leading to a renewal of the doctrine of Scripture, has questioned the foundational place for the doctrine of inspiration as usually expressed establishing both Scripture's identity and authority. Rather he poses that "the concept of inspiration serves to *explain* the identification of the Bible as the Word of God, and the Bible is Word of God by virtue of its origin."[64] Inspiration then, as I will argue, remain necessary not as a foundation for understanding the doctrine of Scripture but more rightly as part of Pneumatology and anthropology.

Divine Agency Develops from Three Factors Commonly Understood in Early Modernity and the Possibility of an Alternative

As indicated earlier, it will be argued that the formation of an understanding of divine agency in the world that developed with a conjunction of three factors in early modernity. These three factors are how the divine perfections were understood, the application of the notion of the two books of God's revelation in Scripture and nature, the broader use of the idea of divine agency contained in the doctrine of inspiration. This particular understanding of divine agency relies on the existence of the soul as an essential component of human anatomy and that God's action within the soul or in similar manner in the world is supposed as perfect. If a Christian description of divine agency must continue to rely on either assumption, then logically, faith stands on an all-or-nothing basis depending on demonstrable proof of the perfection of God's action in nature.

As this logical connection leads to an unsustainable conclusion it would be tempting to abandon inspiration. Unfortunately, such abandonment discards the reality of personal contact between God and humans and abandons an essential element of historical Christianity. If it can be shown that the problematic logic is merely a conclusion derived from a flawed description of inspiration, then the dichotomy is false and such personal contact need not be forsaken. If as it will be argued that the agency and action of God do not automatically imply perfection by human standards, then what does happen when God acts in nature or through a human being? The contention here is that the understanding of divine agency which developed from use of an inadequate description of inspiration when combined with non-christological understandings of divine perfections and the notion of

---

64. Watson, "Hermeneutics and the Doctrine of Scripture," 9n20.

the two books forced an all-or-nothing dichotomy between divine perfection and divine non-existence. This impasse can impede the resolution internal to belief of dialogue between theology and science. This is arguably demonstrable, for example, in the writings of Darwin and Huxley. If an individual lacks the ability to resolve such an internal dichotomy or simply lacks confidence that it can be resolved it may well be expected to affect how they engage in the wider debate between theology and science.

This book will trace how this dichotomy develops. The next chapter will review the first two established factors and put the case for the importance of inspiration as a third. It will also outline the detail of inspiration's dominant Augustinian expression. While the next chapter will describe the understanding of these three factors in early modernity, those that follow will describe how the understanding of divine agency develops.

One important aspect of the next chapter will be to explore how the doctrine of inspiration developed in the west. Augustine's description has its roots in Tertullian's and be shown to draw on Aristotelian and classical medical ideas including those of the philosopher Cleanthes and the gynaecologist Soranus of Ephesus. In the Augustinian description of inspiration, the soul is understood to be a metaphysical element of a human which is necessarily stood aside during the direct action of the Holy Spirit. Therefore, *ekstasis* is automatic when the Holy Spirit acts. This development has not been previously thoroughly explored. Most relatively recent treatments of the development of the doctrine of inspiration deal solely with inspiration presuming it only pertains to Scripture.[65] None seem to deal with inspiration in broader terms such as that used by Calvin's in *Institutes* of there being related divine and secret inspiration. While Augustine's description becomes dominant, his bitter controversy with Jerome suggests the existence of a different but neglected understanding of inspiration. It is possible to trace a differing line of theological argument which will be developed as an incarnational description of inspiration. This will open up a different way to understand divine agency in humans.

As will be described in the next chapter it is only after the Reformation that the Augustinian description's emphasis on the perfection of God's action in communicating Scripture combines with a more general understanding of the divine perfections to bear the weight of scripture's authority as Protestants acted to exclude the suspect authority of the church. In the seventeenth century there was a renewed application of the notion of the two books of God's revelation to the understanding of the natural world.

---

65. Benoit, *Revelation and Inspiration*; Gaussen, *Thoepneustia: The Plenary Inspiration of the Holy Scriptures*; Marshall, *Biblical Inspiration*; Sanday, *Inspiration*; Sasse, "Inspiration and Inerrancy"; Sasse, "Concerning the Nature of Inspiration."

Also there was also a resurgence of interest in Augustinian anthropology among Newton's contemporaries that shaped the development of natural philosophy.[66] The third chapter will demonstrate the influence of the Augustinian description of inspiration in the development of Natural philosophy in the seventeenth century in the work of Isaac Newton. This will highlight a common theological dimension throughout Newton's work which has previously been overlooked and will address some aspects of his thought which are known not to be fully explained by existing descriptions of his work. It will be shown that Newton's understanding of the spiritual nature of matter allowed him to postulate the action of God's omnipotence inspiring the natural world in a manner that parallels the Augustinian description of how the Holy Spirit's acts during inspiration of humans.

Newton's notions retained influence in various aspects of natural philosophy and natural theology during the next century. However, what is more significant for the debate between theology and science is that Newton's successes fuelled the assumption that it was possible to discover God's communication written in the world's natural order unalloyed by the fall or the taint of sin. As a corollary of perfect divine inspiration, this led to presuming that what God has chosen to communicate will have been perfectly recorded in either Scripture or nature. By the turn of the nineteenth century, detailed examination of the book of nature yielded a mounting body of evidence that did not meet this expectation that God's communication would be revealed in its perfection.

Therefore the fourth chapter of this book demonstrates through the thought of Darwin and Huxley how the notion of the two books of revelation and its integral metaphysical Augustinian description of inspiration came to be at odds with what was being discovered of the world and how it led both to a studied place of agnostic uncertainty concerning God, the soul and the possibility of God's communication through the soul or the world. Because both assumed perfect divine action and metaphysics do not hold, this had led to questioning the reality of God's personal contact with humans. For this reason the theological task becomes one of offering an account of inspiration which is not linked to perfection or to an Aristotelian metaphysical anatomy.

Having demonstrated that the description of divine agency contained in the metaphysical Augustinian description of inspiration has been at first influential in the development of natural philosophy and later poses problems for science, it remains in the last part to establish whether the posed incarnational description overcomes these problems. This proposed

---

66. Harrison, *The Fall of Man and the Foundation of Science*.

incarnational description of divine agency is based in the theological notions of *anhypostasia* and *enhypostasia*. In this description, inspiration as the Holy Spirit's action in humans is seen to derive from the unique action of the Holy Spirit in the humanity of Christ. This revised incarnational basis for divine agency in humans makes no assumptions about perfection or metaphysics.

The proposed incarnational description of divine agency will undergo examination in conversation with the Pneumatology of Karl Barth in order to establish that it is worth serious consideration. Barth's incarnational or christological Pneumatology offers an account of the Holy Spirit's work that is not wedded to perfection or metaphysical assumptions. The extensive nature of Barth's theology allows the study of inspiration in relation to broader academic concerns. These concerns include those underlying the development of the dialogue between theology and science: theological understanding of creation and anthropology; consistency in Pneumatology; and the doctrine of Scripture.

This proposal warrants serious consideration if it is theologically coherent and remains plausible while resolving or avoiding a range of known difficulties. The last section establishes this coherence and plausibility. As a revised description of inspiration, this proposal will be shown to avoid past problems, while being able to stand in engagement with a breadth of theological issues. Such a revision of inspiration is one necessary step of many needed to resolve underlying issues in the breadth of the dialogue. Doubtless, inspiration is not the only reason for underlying unresolved issues between theology and science. It is the aim of this work to provide one building block needed for enabling conciliatory debate between theology and science to continue.

CHAPTER 2

Divine Agency, Inspiration, Perfection, and Generic Theology

THIS CHAPTER TRACES HOW three factors converge in the early modern period leading to the development of a description of divine agency that has, largely uncritically, been assumed in and shaped the dialogue between theology and science. These factors are:

1. The theological application of God's action as a perfect being to the natural order;
2. The application of the notion of the two books of God's revelation (both those of Scripture and nature); and
3. The commonly understood doctrine of inspiration becoming the arbiter of divine authority in protestant thought.

Before exploring the third factor in detail I will firstly review established treatments of the first two. I will note, particularly for the first, that conflict and concurrence between differing views continue to exist in contemporary scholarship. The third factor is the largely unexplored role of how the doctrine of inspiration, when added to the mix, led to the development of a generic understanding of divine agency in the world. I argue that the doctrine of inspiration contained an understanding of divine agency in human beings that became the basis in early modernity for the development of a description of divine agency in the world. The last section of this chapter will offer an alternative description that avoids the particular problems that have led to the particular unresolved underlying issue that is the focus of this work.

While not the focus of this book, it is necessary to understand how the nature of God's perfections were understood in order to appreciate its influence on the development of the description of divine agency and how that has affected the relationship between theology and science. While understanding of divine perfection was not altogether static from the seventeenth to the nineteenth centuries, many of the questions raised in its practical application remained unresolved throughout these centuries. Indeed debate continues,[1] making this question another unresolved underlying issue for the dialogue between theology and science.

The second factor, the notion of the two books of God's law became an interpretive tool for understanding the world. It has been argued extensively that the methodologies developed for understanding the book of divine revelation were adapted with similar rigor to investigating the book of nature.[2] This rigor of interpretations explains much of the impetus given to the disciplined study of nature. Nonetheless, it will be argued that in itself this does not explain where the understanding of divine agency that developed in early modernity came from.

This understanding comes from the previously unconsidered third factor, which is implicit in early modernity's development of the doctrine of inspiration. The manner by which God was assumed to interact with the metaphysical soul in inspiration came to complement the development of generic description of the divine perfections and the notion of the two books as it was being applied to the study of nature to develop an understanding of divine agency. It is first necessary to explore the shape of the doctrine of inspiration used by the early moderns, before exploring the shape of divine agency within humans contained within the description. This will lay the foundation for further exploration of historical developments in the modernity's dialogue between theology and science. It is this understanding of divine agency as applied to scripture which meant that inspiration had become, in at least the particular usage amongst Protestants in the seventeenth century, the guarantee of God's flawless communication to humanity unstained by human frailty, unholiness and limitations. It is argued that all three factors interact to give rise to a particular description of divine agency which similarly guaranteed God's purposes in nature in spite of its frailty and flaws. The way this developed was by presuming an impersonal generic understanding of God with a reliance on a particular way of God interacting

---

1. Barth, CD II/1:4; Rogers, *Perfect Being Theology*; Clayton, *The Problem of God in Modern Thought*; Plantinga, *Where the Conflict Really Lies*, 50–100.

2. Brooke, *Science and Religion*; Brooke and Cantor, *Reconstructing Nature*; Brooke, "Reading the Book of Nature"; Harrison, *The Bible, Protestantism and the Rise of Natural Science*; Harrison, "The Book of Nature and Early Modern Science."

with a particular understanding of human anatomy which thus carried with it the seeds of later problems. What will remain unexplored until the next chapter is the application or extension of this description of divine agency in humans to a related divine agency in the world.

The Convergence of Augustinian Inspiration and Perfect-being Theology in Early Modern Science's Synthesis of an Impersonal Description of Divine Agency in the World

The doctrine of inspiration as used in Newton's time (late seventeenth and early eighteenth centuries) came to be added to prior understandings of divine perfection and the two books leading to developing a description of divine action in the world. This form of divine agency was described independently of Trinitarian theology and also Christology. It is a generic description which gives an account of divine action without reference to who God is. This omission of the centrality of the incarnation is striking. This description uses the elements of Augustine's description of inspiration (which in turn will be demonstrated to be based on Tertullian's) which requires the existence of a metaphysical soul and the assumption that unalloyed divine action in the world is perfect. This usage runs contrary to both Augustine and Tertullian who never wrote of divine agency in the world without reference to God's action in the world in Christ. Nevertheless, and unusually for their theology, their work on this issue easily permits such interpretation: their description of divine inspiration will be also shown to be based more on philosophy and medicine than Scripture.

What developed was an understanding of divine agency in the world which was in turn based in God's perfect attributes and also derived from the way God was understood to interact with the human soul. This application of Augustinian inspiration left two broad questions for the dialogue between theology and science. The first which became important over the eighteenth and nineteenth centuries was to determine how God's action in the world could be spoken of as perfect. What becomes apparent that there is movement from a blunt ideal perfection to various forms of teleological perfection in which the bounds of speaking of such perfection became more and more restricted. In a more mature form this notion of teleological perfection was specifically rejected by Darwin. Rogers[3] and Clayton[4] demonstrate that the issues that developed concerning divine perfection remain

---

3. Rogers, *Perfect Being Theology*.
4. Clayton, *The Problem of God*.

a subject of current debate. While Rogers seeks a ways to reclaim perfect-being theology, Clayton argues for a solution to these unresolved issues by using process theology, by beginning with an affirmation of the personhood of God. Neither Rogers nor Clayton, however, address the inherent problems that Barth identifies that arise from their choice to begin the process of analysing the divine perfections with abstract, thus impersonal and generalised, understandings of the divine freedoms.[5] The similarities and differences between Rogers, Clayton and Barth will be discussed more fully in the next section of this chapter as they both explore and are representative of the breadth of discussion of divine perfection. What is primarily relevant to this discussion is that perfect-being theology was held during early modernity to be an accurate way to describe God by theologians and natural philosophers as they sought to understand how the world works and how God is related to the world. Barth actively argues against beginning with the perfections of God's freedoms. Briefly, Barth's extensive argument may be summarized as, if one begins with any abstract discussion of God's freedoms inevitably leads to an impasse between speaking of the perfection of divine agency and the possibility of God's divinity or existence.[6] Instead, Barth argues that it is better to begin with the perfections of God's love, demonstrated christologically. Such an impasse can be avoided by beginning with whom God has shown God's self to be, rather than by considering what God might be conceived to be.

The question of how to describe divine agency in the world can be considered a particular case in which impersonal understandings of God's perfections have led in application to the description causing its own setbacks. The solution which will be offered as an alternative will recast divine agency in christological/pneumatological terms and shows that such a description does not lead to the same problems as the understanding of divine agency that actually developed historically. However, such an alternative cannot stand unless the following question is also addressed: Why was a conceptual leap made between understanding divine agency in human beings to divine agency in the world in general? Why as there is no particular reason for these to be closely linked? Nonetheless, in the development of the relationship between theology and science, the possibility of divine agency in the world has been assumed to be predicated by the necessity of the existence of the soul and God's interaction within it. The soul's existence was questioned in debate between theology and science by Huxley who held that unanswerable questions about the existence of the soul implied a similar inability to

---

5. CD II/1–CD IV, *passim*.
6. CD II/1:338–41.

resolve the question of God's existence.[7] The question of the soul's existence continues to be a source of contemporary debate.[8]

It is intended to demonstrate that a christological/pneumatological or incarnational description of divine agency in human beings can be developed which does not depend on the existence of the soul or any presumption about the anatomical or metaphysical makeup of humans.

The remainder of this chapter will review the three factors which it is claimed contribute to the early modern understanding of divine agency, in particular how these factors were widely held during the seventeenth to nineteenth centuries. In doing so the discussion will explore five subjects which are relevant to following chapters.

The first subject is how the issues involved in the medieval discussion of divine perfections continue to be relevant in contemporary debate. This is in spite of questions raised during the period marked by development of what has become modern science. It will be suggested that some of the issues which remain in present debate were also relevant at times throughout the last three centuries. The second subject is how the nature of divine perfection was understood in the early modern period; in particular those aspects which would have been relevant to constructing a generic understanding of divine action in the seventeenth century. The third subject is how understanding of the natural order began to be shaped by new understanding of the bible as one of two books of God's revelation. The fourth subject is how inspiration came among Protestants to bear the weight of guaranteeing the authority of Scripture and how this came play a similar role in understanding creation. Then the final subject will be to explore the nature of the doctrine of inspiration exclusively used in the seventeenth century and the understanding of divine agency which was encapsulated within this Augustinian description. Whilst this description was exclusively used, nonetheless, a detailed analysis will demonstrate that it is not the only possible description. At this point it will be appropriate to offer and contrast the alternative incarnational description that will be explored later.

This discussion will provide a necessary background for later chapters which will explore how Newton develops his understanding of divine agency based on the doctrine of inspiration; how Darwin and Huxley become agnostic about the possibility of divine agency; and determining whether the incarnational model for divine agency is theologically coherent and remains plausible while resolving or avoiding a range of known difficulties.

---

7. Huxley, "Has a Frog a Soul?"; Huxley, "On Sensation and the Unity of Structure of Sensiferous Organs"; Huxley, Agnosticism and Christianity, *Collected Essays*,

8. Russell et al., *Neuroscience and the Person: Scientific Perspectives on Divine Action*.

Perfect-being Theology Remains a Contemporary Issue[9]

Rogers and Clayton explore divine perfection and divine agency as an ongoing problem. Both speak of resolving this problem in different ways. Rogers tries to reclaim the ideal perfection of God's freedom to act omnipotently, with omniscience, and omnipresence through the holistic reapplication of medieval philosophical methods. Clayton, on the other hand, attempts to preserve the personal nature of God and consequently argues that limitations are required to traditional understandings of God's perfections. These revisions he bases in process theology. Writing in the early twentieth century Barth offers a third alternative addressed by neither Clayton nor Rogers. Barth argues, in general, that approaches as typified by Rogers and Clayton will inevitably lead to insoluble tensions. Nonetheless, all three generally agree about how the understanding of God's perfections developed from the late Middle Ages to the present. Thus it is useful to bring these three into dialogue in order to describe what understanding of divine perfection underlay the development of the dialogue between theology and science.

Perfect-being Theology: for and against

In relation to the contemporary question of divine perfection, Rogers notes that

> there is a pervasive tension in Christian thought between "the God of the philosophers and the God of the Bible," between God as "wholly other" and God as a partner in interpersonal relationships, between God as the absolute, ultimate source of all being and God as the dominant actor on the stage of history.[10]

This tension demonstrably develops during the late eighteenth century to develop in full flower in the latter half of the nineteenth century, as will be dealt with in the succeeding chapters. Nonetheless, in spite of the tension which has developed, Rogers is able to acknowledge that even in the present,

> [e]verybody seems willing to allow that God is knowledgeable, powerful, good and a free agent. With respect to these attributes the questions that will arise will concern their scope and nature. But there are other attributes where the issue will be whether

---

9. For an expanded version of this discussion, see Brennan, "On Why We Should Agree with Contemporary Atheists—Or Why a Generic God Does Not Exist."

10. Rogers, *Perfect Being Theology*, 9.

or not they should be considered "great-making" properties at all. There are those who do not see any value to the traditional attributes of divine unity, eternity and immutability. Some, most notably process theologians . . . see multiplicity, temporality and changeableness as virtues, inextricably bound up with the concept of a God who is really related to His creatures.[11]

While arguing for perfect-being theology in spite of its problems in application, Rogers concedes, "It could be insisted that we must accept a God who is so radically limited since this is the only way to resolve the various puzzles raised by the concept of a perfect being."[12] She is not content to leave matters stand thus, asserting "that it is possible to resolve the paradoxes and leave God's infinite perfection and absolute sovereignty intact."[13]

Conversely, Clayton does not avoid, but rather embraces the abandonment of theological orthodoxy in favor of a description of God which he sees provides the only possible answer to his dilemma in "understanding divine agency."[14] In order to preserve the possibility of personal interaction between God and the world and leaving open the possibility of similar personal interaction with humans, Clayton abandons the notion of divine transcendence as otherness, except in degree. Clayton argues that it is difficult to see how creation could truly have an existence independent of God. "It is clear that, given the resources of panentheism, a theory of divine agency no longer confronts the problem of absolute differentness. Further, the position sketched here, which links God as a being to the world and yet also to the infinity of God's nature, does offer possibilities for reconciliation that are not open to more traditional positions."[15]

Rogers recognises the motivation behind such challenges to classical understandings of perfect-being theology:

> It is often said that the God of Augustine, Anselm and Aquinas, the immutable, eternal, transcendent source of all, cannot be the personal, loving God who acts as an agent in the world and takes an interest in individuals. Process theologians, preferring the latter image of the divine to the former, take this position to its extreme, thoroughly repudiating traditional perfect being theology.[16]

11. Ibid., 12–13.
12. Ibid., 13.
13. Ibid.
14. Clayton, *The Problem of God*, 505.
15. Ibid.
16 Rogers, *Perfect Being Theology* 8.

Rogers also claims the problem with this form of theology is that it becomes theology which lacks a recognisable continuity with that of the biblical and historical faith. Rogers' revision of perfect-being theology claims that contemporary analytic methods tend to obscure an appreciation of the forest by having too narrow a focus on the trees. This narrow focus makes description of the traditional attributes of God seem "confusing and inconsistent."[17] Rogers argues that certain medieval methods of epistemic and metaphysical analysis are more holistic and have not been, as yet, exhaustively explored in describing God's perfect attributes.[18] Rogers argues that these notions need deeper and more profound reassessment of those attributes in the light of modern physics and science.[19]

Clayton, however, claims that there are commonly held weaknesses of traditional perfect-being theology and that these weaknesses contribute to what has been an ongoing problem of how to best speak of the nature of God's perfections.[20]

> Inconsistencies have been alleged between almost all the "perfect-making properties" even by perfect-being theologians: between immutability and omniscience and omnipresence, between omnipotence and impeccability, between omniscience and immutability, between omnipotence and freedom.[21]

Clayton contradicts Rogers' solution arguing that the notion of perfection is not viable. Clayton, while concerned about the rationality of God talk, remains optimistic that God talk is "an object of direct rational inquiry"[22] and also that it is possible to develop this as "an instance of a regulative principle that can also be evaluated as a constitutive theory."[23] He is, however, pessimistic about continuing to be able to refer to God is terms of abstract infinite perfection—in a universally accepted epistemically perfect metaphysics—"since by definition no one can know what lies beyond all possible knowledge."[24] Clayton asserts that some basic questions have not been asked: "Is the notion of a perfect being coherent, or does it . . . depend on assumptions we can no longer make?"[25] Unlike Rogers, Clayton

---

17. Ibid., 6.
18. Ibid., 8.
19. Ibid., 108.
20. Ibid., 154.
21. Ibid., 134–35.
22. Ibid., 40.
23. Ibid.
24. Ibid., 36.
25. Ibid., 133–34.

INSPIRATION, PERFECTION, AND GENERIC THEOLOGY   31

believes that notions of perfection do stand on such assumptions that can be no longer made and that when drawn to their logical conclusions they become incoherent.

Like Rogers, Clayton sees the ground of perfect-being theology lying in Anslem's ontological proof that "the intuition of perfection gives rise to an exact and detailed analytic discussion of God's attributes."[26] However, Clayton argues "[p]erfect-being theology looks rather different if one is not convinced of the soundness of the ontological argument as a proof."[27] Clayton's argument focuses on the nature of divine agency in the world. He begins with God's supposed perfection and demonstrates repeatedly how difficult it is to conceptualize such action without it in some way limiting God's freedom to act or choose and thus limiting God's perfection.[28] If God's perfection is considered of necessity to encompass God's freedom to act without external restriction, then the choice seems to lie between redefining how we understand that perfection in the world and choosing to radically revise how God might be described.

Nevertheless, Rogers' offers a number of salient points for the present discussion. She offers a description of how at the beginnings of modernity God's perfection was understood to be reflected in creation. While various questions have arisen regarding the applicability of perfect-being theology, it continues to be used. Indeed, Rogers by presenting evidence of a larger debate attempting to understand God as perfect-being shows perfect-being theology retains currency in contemporary discussion.

Barth Questions Both Approaches

What both of the preceding approaches to divine perfection have in common is that they are non-specific, attempting to talk about God in generalized terminology. Barth, in effect, argues that such attempts and in particular any attempt to speak of God's perfections generically will lead to the kinds of problems that Rogers identifies with process theology and as well as those that Clayton identifies with perfect-being theology. Barth gives four examples which he claims have in common the essential problem of all such generic theologies: They treat God firstly as an object that can be known in terms of preconceived generalized ideas rather than beginning with the particular and unique revelation of God.[29]

26. Ibid., 132.
27. Ibid., 134.
28. Ibid., 504.
29. CD II/1:338; CD II/1:339; CD II/1:339–40; CD II/1:240–341.

Barth notes that where distinction has been made between the various perfections of God, nearly all of such discussions begin with what God is rather then who God might be. Barth identifies that the supreme difficulty is assuming that these attributes can be known generically in and of themselves and then applied to God.

> In this connexion we may consider as obvious errors all those types of a doctrine of attributes which attempt to define and order the perfections of God as though they were the various predicates of a kind of general being presupposed as known already, whereas in reality each of them is the characteristic being of God Himself as He discloses Himself in His revelation. The right way, on the contrary, will consist in understanding the attributes of God as those of this His special being itself and therefore of His life, of His love in freedom.[30]

Rather than move from a general idea to a specific, Barth argues that the reverse needs to be the case when speaking of God, i.e. one must first deal with the specificity of God revealed in the incarnation of Jesus Christ.[31]

Also, Barth observes, starting with the "what" makes the large assumption that the nature of the perfections of God can be known without firstly considering the extent to which God has made them knowable. Barth remarks that we can only know the perfections of God in as much as and only in as much as God makes those perfections known. For example the question, "Can God make an object too heavy for God to move?"—is unanswerable except if God chooses to reveal enough to make such an answer known.

Barth, in particular, argues that by beginning to talk about God's perfections by beginning with (the perfection of God's freedoms such as) omnipotence, omniscience, eternity and immutability is to assume "that God is first and properly the impersonal absolute."[32] This risks not apprehending the very object of the investigation.[33] To begin thus "corresponds neither to the order of revelation nor to the nature of the being of God as known in His revelation."[34] These have, he argues, a particular order which is freely imposed on God solely by God's own free choice. While a distinction must be made between God's "sovereign freedom and the perfections proper to it, eternity, omnipotence and so on"[35] and "the love of God and its perfec-

30. CD II/1:337.
31. CD II/1:334.
32. CD II/1:349.
33. CD II/1:348.
34. CD II/1:349.
35. CD II/1:345.

INSPIRATION, PERFECTION, AND GENERIC THEOLOGY

tions, holiness, justice, mercy and so on,"[36] the latter cannot be treated as secondary. Barth puts these first, starting with the perfections of God's love demonstrated specifically in the grace God has demonstrated in Christ, beginning with mercy and righteousness then providence and wisdom.[37] Only then does he seek to understand the perfections of the divine freedoms.

> This way can consist only in our thinking first of the love of God as it really exists in His freedom and then of His freedom as it really exists in His loving. But the "first" and "then," the sequence can be reversed only arbitrarily and at the cost of great artificiality and misapprehension. We cannot allow ourselves such caprice. Therefore we begin with the perfections of the divine love: with the intention and in the confidence that in this way, even if indirectly, we are beginning also with the divine freedom.[38]

Barth claims the change in order is essential if the problems of earlier theologies of perfection are to be avoided. The freedoms of God to act or not act cease to be predicated on any given set or ordering of absolute ideas but on God's gracious prior choice to order who and how God has chosen to be God.[39] In casting the discussion of God's perfection in this way Barth makes consideration of the interactions between God and the world or God and people both personal and particular. It cannot be, Barth argues, considered generically. Such generic descriptions are from the outset false systemisations which obscure rather than highlight the nature of God's perfection expressed in God's power and freedoms.

Because Barth's description of omnipotence begins with God's loving grace, Barth's first issue with divine omnipotence is not evil but sin. "[N]othing of all that He does or is as such should be regarded by us as less real because it is in fact in the new relationship that He does it, as the Creator and Lord of His fallen creation."[40] God's power to act and God's immutability are constrained by God's own choice to perfectly love in the manner made known in Christ. God's first focus is to overcome sin by love. This christological expression of God's perfect love also shapes Barth's description of God's omnipresence. The shape of the distinction between heaven and earth is defined personally and particularly. The interaction of God with the world is not expressed in an idea or a principle but in the person of

---

36. Ibid.
37. CD II/1:368–39.
38. CD II/1:352.
39. CD II/1:353.
40. CD II/1:506.

Jesus Christ.[41] God's perfection can only be known in and through this same person primarily expressed in the perfection of God's love expressed to and recognized by actual people.[42] Hence, Barth concludes in relation to God's relationship with nature:

> We cannot hold, therefore, that there is first a divine power generally in nature, and then in the whole of the course of history, which can be identified in an undefined and rather uncertain way with the power of God, but that this is a matter more of supposition and inkling than of knowledge, perhaps the recognition, in common with heathen religions, of the substance of a neutral supreme power and activity before which, as an obscure but true revelation of divine omnipotence, we can stand reverently for a moment (and at a pinch a little longer) before going on to God's special, concrete, historical capacity and activity and therefore to the true and clear knowledge of God.[43]

What Barth asserts is that particularity is needed. This runs counter to the generic tone of theology's interaction with natural philosophy throughout seventeenth to nineteenth centuries. What is implicit in Barth is that theology can only truly engage with science if it recognises the particularity and personal nature of God's interaction with humanity and the world.

To summarize Barth's points one could say that any generic description of divine agency would be predicted, in the face of contradictory evidence, to develop into an unresolvable choice between finding new ways to assert absolute truths or to abandon traditional ways of speaking of God. Conversely, describing divine agency while first taking seriously the particularity of a christological understanding of God's interaction with the world may avoid such an impasse. Rogers and Clayton are examples of divergent contemporary opinions regarding perfect-being theology which serve to illustrate this point as they deal with divine perfections as abstractions. They also indicate the difficulty in resolving the question of perfect-being theology. It will be argued that resolving issues related to divine agency need not depend on resolving the question of perfect-being theology as Clayton assumes.

---

41. CD II/1:478; CD II/1:486.
42. CD II/1:599.
43. CD II/1:602.

Perfect-being Theology at the Beginnings of the Modern Period until the Mid-nineteenth Century

While Rogers, Clayton and Barth differ greatly on how to resolve issues related to describing divine perfection, they largely agree about the shape of perfect-being theology during the beginnings of what later becomes modern science. The ongoing nature of problems and unresolved issues involved in speaking of God's perfections has been noted. It is now useful to describe how God's perfections were comprehended as one of the three factors that gave rise to the seventeenth-century natural philosophers' understanding of divine agency. Changes in the comprehension of God's perfections developed slowly making this a relatively stable factor among the natural theologians of the eighteenth century. While there is little disagreement that this factor contributes to how divine agency was understood, it will be shown that this factor is not sufficient in itself to explain how the understanding of divine agency evolved.

Rogers describes the basis for discussions of God's perfections in the medieval notion of divine simplicity, drawing on the essential elements from Anselm, Augustine, and Aquinas. Clayton also shows that these ideas are mirrored in Déscartes' thought.[44] The basic argument goes thus: "a perfect being must be unlimited, but to be composite or complex is to be limited."[45] God is act and not "inert property"[46] Thus, "when Aquinas (or, I take it, Anselm or Augustine) says that God is His omnipotence, omniscience, omnibenevolence and so on and that all of these are identical with God, what [h]e means is that God is His act of knowing and doing and being perfectly good and these are all one act."[47] God knows and relates and creates and is—in one action. An essential aspect of this application of perfect-being theology was the linking of infinity to perfection. That is infinity as unlimited, rather than the more commonly used medieval notion of infinity as incomplete. [48]This was in order that there be no limit to God's perfection. Déscartes follows a "general principle, namely that the infinity argument should not induce one to ascribe to God anything incompatible with divine perfection."[49] God's actions must reflect God's unlimited perfection, thus creation as one of God's actions must in some way reflect God's unlimited

---

44. Clayton, *The Problem of God* 132.
45. Rogers, *Perfect Being Theology* 25.
46. Ibid., 29.
47. Ibid.
48. Clayton, *The Problem of God* 177–82; Rogers, *Perfect Being Theology* 12.
49. Clayton, *The Problem of God*, 156.

perfection. However, creation cannot be as perfect as God. Nature's reflection of God's perfection was understood as carried by participating in the being of God as the source of its existence.

> [P]erfection and participation belong indissolubly together. If the highest being (or being itself) is viewed as perfect, all other beings will have their being, and whatever other positive qualities they have, only through their participation in the highest instance. . . . Participation characterizes the relation between the most perfect and the less perfect. . . . the more perfect bestows and the less perfect receives a perfection. Consequently, there is always an ontological priority of the more perfect in participation.[50]

Clayton also notes that theologians since Augustine have appealed to the doctrine of creation out of nothing. "[P]articipation becomes, it was thought, a happy way of expressing the ontological analogy between creation and creator. The goods or perfections of created beings are then similar to their source because they have these perfections through participation."[51] The perfections of this world can be described as out workings of God's perfection and God's choice to act in the world. As Rogers details,

> Augustine in *On the Literal Meaning of Genesis* . . . explains that, "if He is not able to make good things then He has no power, and if He is able and does not make them, great is His envy. So because He is omnipotent and good He made all things very good." Given God's nature He could not do other than He does.[52]

It follows that the world is not only good but that it is the best possible world. Clayton claims this notion reached its highest expression in the work of Leibniz. Rogers agrees the idea of a best actualisable world coheres with how divine perfections were understood in early modernity. "Such a universe would be the best possible reflection of divine infinity, but could never equal it, since the creature is necessarily on an ontological level radically inferior to the creator."[53] Further, Rogers indicates that it was generally held that the world must also be the best possible world as an outworking of divine omnipotence. "'Could God have done other than He does?' . . . Augustine is clear that God, while absolutely free in the sense of acting purely from His own will, inevitably does the best. The Supreme Good is necessar-

---

50. Ibid., 163.
51. Ibid., 168.
52. Rogers, *Perfect Being Theology*, 34.
53. Ibid., 104.

ily self-diffusive, and so God could not fail to create."[54] This was not the only opinion and Rogers notes that Aquinas adopted a rather different position.[55] In either case creation is to be understood to be as perfect as it could be. Rogers summarises the similarity of the conclusions of these differing approaches as "God takes our free choices into account and actualizes the best world compossible with these choices."[56] She notes that the notion that the world is the best it could be continues to be "reflected in the contemporary discussion."[57] "Compossible" meaning "able to coexist" was a term coined by Leibniz. Rogers avers that there is a problem with the notion that this world is the best of all possible worlds, in that the nature of suffering and evil commonly leads "contemporary philosophers to hold that God must be free in a libertarian sense and be able to choose evil, or at least do more or less good, so that He can get credit for doing well when He could have failed to do so.[58] Sin, suffering and evil seem to put this notion of God's perfect action to the lie. Rogers acknowledges in these views "the problem of evil becomes acute, and it is very difficult to absolve God of the responsibility for the sin and suffering in this world."[59] But this did not stop late medievals and early moderns from affirming that God's action in the world must be perfect.

Clayton explains the logic of the argument as follows. "clear bifurcation emerges between a most perfect being and those things that are less perfect (say, the created order)." There are qualitative and quantitative differences between the divine and human levels: Qualitative, "because one level is infinitely beyond the other"[60]; Quantitative, "because the lower has its existence from the upper (whether through creation or emanation), has some knowledge of it (if only the knowledge that it is infinite), and should strive to understand and to emulate it."[61] Clayton notes that there remain problems with this logic.

> For example, it seems to imply that the created order is *as such* imperfect, which clashes with the teaching of the basic goodness of creation in the Christian tradition. Still, it does seem that the monotheism of Jews, Christians, and Muslims requires that God be perfect, the highest moral instance, and the cause of goodness

54. Rogers, *Perfect Being Theology*, 102.
55. Ibid., 103.
56. Ibid.
57. Ibid.
58. Ibid.
59. Ibid., 35.
60. Clayton, *The Problem of God*, 167.
61. Ibid.

in the world—however perfection is to be understood. It could well be that the two-category account of divine perfection will survive the serious difficulties that we shall soon encounter with more lofty forms of perfect-being theology.[62]

In spite of these inconsistencies, this was the state of perfect-being theology at the turn of the eighteenth century. The loftier form of perfect-being theology to which Clayton refers is that expressed in Leibniz's writings. Clayton demonstrates that it is Leibniz's notion that God's perfect action is expressed in creation's purpose which causes it to strive to understand and to emulate God's perfection. This teleology develops in importance as the description of divine agency evolves through the eighteenth century. It remains important well into the nineteenth century.

Leibniz on the Perfect Harmony of Nature

Leibniz developed the notion of the world being the best of possible worlds as that which best achieves God's perfect ends. Clayton sees Leibniz as giving the most refined expression to perfect-being theology in early modernity.

> With Leibniz we jump directly to the apex of this tradition in early modern philosophy. Here is a thinker who, working with largely Cartesian assumptions, became the last major modern philosopher to build a distinctive system around the *ens perfectissimum* as its core. By contrast, the infinity of God now plays a slightly less important role as a complement to the perfection of God.[63]

Leibniz commented toward the end of his life:

> No more am I able to approve of the opinion of certain modern writers who boldly maintain that that which God has made is not perfect in the highest degree, and that he might have done better. . . . I think that one acts imperfectly if he acts with less perfection than he is capable of. To show that an architect could have done better is to find fault with his work.[64]

---

62. Ibid.
63. Ibid., 183.
64. Leibniz, "Discourse on Metaphysics," Par. 3, 5–6.

INSPIRATION, PERFECTION, AND GENERIC THEOLOGY   39

Comparative perfection is insufficient, noting Leibniz's further qualification that God's action would be perfect "in whatever way God had accomplished his work."[65] Clayton offers the argument that

> the principle of perfection provides one of the central organizing pillars, and arguably the most central one, for Leibniz's system. The perfection idea runs through his natural philosophy, his practical reflection, and of course his theology; it is also central to any knowledge of God.[66]

Clayton demonstrates that Leibniz's understanding of God's agency in the world is characterized by application of divine perfection, his principle of sufficient reason, and teleology so that the world is considered as being the harmonious out-working of divine prefect action. Nonetheless, he is concerned to clearly restate Leibniz's position carefully. Claiming that most surveys of Leibniz,

> often pay lip service to the role of perfection, though its systematic function as an organizing principle is, I think, usually underappreciated. Perhaps more important, the systematic consequences of the weaknesses of this idea in Leibniz have rarely been treated.[67]

While carefully and critically restating Leibniz, Clayton is, of course, unsatisfied that Leibniz's arguments continue to be persuasive. Clayton summarises his broad ranging analysis of Leibniz's large and varied corpus in relation to the relationship of perfection to the natural sciences by asserting that

> Leibniz's overarching goal was to provide a general account that allows for both mathematical and mechanistic explanations of natural events, while at the same time leaving room for an explanation of the universe in terms of final causes (viz. the goals of God and other monads.) Since metaphysics must supply the overarching framework, this means ideally that physics could derive from the notion of a perfect being.[68]

Leibniz postulates that this world is the best of all possible worlds,[69] in such a way that the world has been created so that free will and God's

---

65. Ibid.
66. Clayton, *The Problem of God*, 202.
67. Ibid., 202.
68. Ibid., 214.
69. CD III/1:394.

purposes are compossible. "[W]e may say that in whatever manner God might have created the world, it would always have been regular and in a certain order. God, however, has chosen the most perfect, that is to say the one which is at the same time the simplest in hypotheses and the richest in phenomena."[70] While Clayton does not believe that Leibniz succeeds, he does conclude that Leibniz believed that God's perfections are worked out in creation in a number of significant principles.

> Leibniz did not succeed in linking natural science closely to his philosophical theology in the manner he hoped for. God's existence may be a sufficient condition for concluding, for example, that simpler hypotheses are preferable (the parsimony principle), but it is not really necessary to call on the authority of theology for defending the use of such principles in science. Many of his principles—for instance, "natural science should seek to optimize explanations, covering the most data by means of the least complex hypotheses possible"—can just as well be justified on pragmatic grounds, or as regulative maxims for the practice of science.[71]

These are significant in relation to developments in the relationship between theology and science into the nineteenth century. Explanations which are both far reaching and simple or even elegant are to be preferred as they better reflect the perfection of the creator, according to Leibniz. Clayton does not totally reject Leibniz's reasoning. "There are exceptions, of course: postulating God does explain why maxims such as the law of parsimony or the [Principle of Sufficient Reason (PSR)] should be true of the world and not merely crutches that scientists cannot otherwise dispense with."[72] That is that he has difficulty finding a basis in reason alone that should dictate why these principles should be preferred over others. Leibniz's principle of sufficient reason guides much of his work. Clayton observes that

> Leibniz's principle of sufficient reason amounts to the claim that there is a metaphysical reason for all truths and falsehoods. This fact gives it a bidirectional status: for everything that is (exists), there is a reason, and all reasons must (and can) be grounded in the world, the way things are.[73]

---

70. Leibniz, "Discourse on Metaphysics," Par. 6, 10–11.
71. Clayton, *The Problem of God*, 214–15.
72. Ibid., 238.
73. Ibid., 175.

Not only is God's perfection to be understood as reflected in goodness, simplicity and elegance—everything that exists must have a creator-given purpose. Leibniz and his student Wolff were optimistic that these purposes could be understood and actually made God's perfection clearer. Barth comments that Wolff held that "there are various rules affecting world-perfection and therefore its internal coherence . . . A world will be the more perfect the more these compromises actually bring out the underlying rules."[74] There is a demonstrable "harmony in the relationships"[75] between those rules. Even so, some things may remain unknown because the human perspective is too limited.

The question then for Leibniz is what is the role for God's action in the world beyond actual creation and the provision of purpose? Is there a place for divine agency at all? Clayton asks, "What other causal roles does his God play in the world? Barth also describes the Leibnizian philosophy as holding "that God dwells in perfect contentment. Since He has everything to a supreme degree, there remains nothing which He could wish further."[76] God could not wish further or need to act further. The necessity of ongoing divine agency in the world becomes a moot point if the perfection of the world was in any event guaranteed by its clockwork character. Nonetheless, "Leibniz does advocate a role for God in conserving the world, and he follows Déscartes in taking this to involve a sort of continuous creation moment by moment."[77] This conservation of the world is "independent from God. Individual things only need divine assistance in each moment because their being is not contained already in their concepts but is superadded to them by God."[78] "Leibniz is recommending a view of the universe that is teleological as well as mechanical, theological as well as physical."[79] Therefore in application the perfection of God's action must be discernable in the harmonious order of the world. Clayton believes "Leibniz thought that the principle of sufficient reason could accomplish the 'acknowledging of some sort of cosmic orderer'"[80] without recourse to continuous ongoing divine teleological tinkering.

As the perfection of God's action must be reflected in creation's teleology, evil and suffering must logically have a purpose which does not detract

74. CD III/1:393.
75. Ibid.
76. CD III/1:395.
77. Clayton, *The Problem of God*, 238.
78. Ibid., 238–39.
79. Ibid. ,198.
80. Ibid., 199.

from God's perfection. Clayton observes, "[T]he teleological argument also required an explanation for evil (apparent evil) in the world, and thus led directly to the project of Leibniz's *Theodicy*, and hence his ideological argument."[81] However, in, "the *Theodicy* and elsewhere, Leibniz justifies evil in terms of the balance or harmony of the whole."[82] Evil, sin and suffering exist in the world as the minimum necessary to achieve God's purposes. "Hence the best world cannot exist without that which is evil and wicked. God therefore maintains more good by permitting evil and wickedness than by not doing so."[83] Clayton notes that this "is particularly unconvincing to the contemporary reader."[84] The key notion is God's permissiveness. Evil and suffering are not caused by the perfect-being God. The question of evil is linked to the theological question of God's perfection expressed in this way. While this needs acknowledgement as an important issue it is peripheral to the question of divine agency.

The earlier discussion of Barth suggests that these problems with both the nature of evil and God' purposes in creation could be considered as arising from attempting to describe God without reference to who God is. Barth describes a "descending quality of the various presentations"[85] of the nature of divine perfection reflected in nature and evil.[86] This degeneration begins with the "the perfection of the world which Leibniz found in the predetermined harmony of antitheses and steps progressively down:

- to the machine-like character of creation in Wolff, its utility in Lesser";
- to the "God who perfectly wills and creates only the best" ;
- to the "clockmaker who happily invents the best of all mechanisms";
- to the "all-wise Author necessarily to be inferred from the marvels disclosed in the kingdom of ants and bees";
- to the "praiseworthy maximum of all the observable greatness and joy and utility and pleasure of the universe";
- to "the supreme giver of so much cheese, vegetables and root products."

---

81. Ibid., 200.
82. Ibid., 208.
83. CD III/1:395
84. Clayton, *The Problem of God*, 200.
85. CD III/1:396–403.
86. CD III/1:403

INSPIRATION, PERFECTION, AND GENERIC THEOLOGY 43

Barth argues the progression is a train boarded which inevitably reaches a terminus that makes "physical and moral evil"[87] integral to God.

Rogers, Clayton and Barth all share dissatisfaction with how divine perfections were described in the early modernity through to the mid- to late nineteenth century. Nonetheless, what can be discerned are the characteristics of divine perfection as understood by theologians and natural philosophers (and later scientists) which influenced how thought about the nature of divine agency.

Leibniz's highly limited role for ongoing divine agency was not a final word in the overall discussion. There was continued support for ongoing active divine agency. The question of divine agency was considered an essential aspect of Christian faith and continues to be as Clayton notes it is,

> one of the least well-articulated challenges facing theism today. To avoid deism, theists must say that God can be, and is, active in the world subsequent to creation. Yet there are serious problems with maintaining that God continues to intervene directly into the natural order. For example, it would seem to threaten the integrity of this order, disrupting the regularity and predictability of the natural world that is necessary for free and reasonable human action. It also threatens to make a mockery of scientific method, since if there were regular miraculous divine interventions one could never know whether a given natural occurrence even had a natural cause. Finally, divine interventions seem to break the law of the conservation of energy—unless God could act by introducing information into the natural order without any influx of energy.[88]

How divine agency was described and the issues that later developed as a result are the focus of the discussion throughout this work. The preceding discussion highlights that notions surrounding perfect-being theology actually make the question of divine agency problematic. Some other source is needed to fill out early modernity's understanding of divine agency. The contention is that descriptions of divine agency were developed from key characteristics encapsulated within the Augustinian description of inspiration. This both held strengths which helped to advance science as well weakness which later became a source of tensions.

Nonetheless, in relation to divine perfection, Barth observes that the understanding of God's perfections remained unchanged through the early modern period until the mid-nineteenth century. Barth notes that is it to

---

87. CD III/1:404.
88. Clayton, *The Problem of God*, 504.

the credit of "certain German theologians" who broke with the nominalism of Thomism and that of the orthodox Protestant tradition that dealt with the particular attributes of God's perfections. While Barth remained unconvinced that their approach did not fall prey to problems akin to those of the old, what is significant for the discussion that follows in later chapters is that their break with tradition occurs well after the latest of the cases to be discussed (that of Huxley). Rogers, Clayton and Barth all point to the persistence of a view of God's ideal perfection throughout this period in which the question of divine agency developed. Harmonious nature as the outworking of perfect divine action becomes particularly important before Darwin, as this was the way that the hand of God, the generic Providence, was explained and taught in the early nineteenth century.

What has been shown in this section is how the nature of divine perfection was understood in the early modern period. By itself this understanding of divine perfection does not necessarily lead to a particular description of divine agency. In terms of the Leibnizian formulation of the divine perfections, divine agency could even be postulated as superfluous. Nonetheless, such understandings of divine perfections permeated theological thinking along with the other factors to be discussed. Together they were included in attempts to understand the world and God's relationship with it.

The next section will trace how the new understanding of the natural order as one book of God's revelation began to be shaped by understanding of the bible as the other book of God's revelation. Following this, the discussion will explore the nature of the doctrine of inspiration held in the seventeenth century and the understanding of divine agency which was encapsulated in this Augustinian description.

The following chapter will demonstrate how the understanding of perfection and the notion of the two books combine with that of inspiration to constructing a generic understanding of divine action in the seventeenth century.

The combination of these three factors gives rise to an understanding of divine agency in the world which helped to spur on disciplined study of nature while providing an expectation that rational answers are both discernable and knowable by humans undertaking careful study. Nevertheless, the particular confluence of these ideas will also be shown to lead to an impasse between traditional descriptions of these ideas and evidence discerned through the study they helped to promote. What will be proposed is that such an impasse is avoidable by careful revision of the theological description of one of the contributing factors.

It will be possible to demonstrate that early modern science was characterized by people who were trying to describe how God worked in the world by beginning with what God is without referencing who God is.

The Two Books of God's Revelation as a Factor in Early Modern Understanding of Divine Agency

It is important to realize that none of the three factors argued as contributing to the development of the understanding of divine agency operated independently. Notions concerning God's perfections were reflected in scriptural interpretation which is at the heart of the notion of the two books. Both of these factors are reflected in the understanding and application of inspiration.

This section reviews the second of the factors which contributed to the early modern development of divine agency in the world. It will highlight the changes in the understanding of the relationship of nature and God and the change in the understanding of the purpose of nature.

The mediaeval understanding was that nature owed its existence to participation in the being of God. Nature's purpose was to be to sustain and teach humans in their worship of the creator, while nature's workings were a mystery knowable only to God. After development and application of the notion of the two books of God's revelation, nature came eventually to be seen as an additional source of understanding God, but whose propose and workings might be disclosed through careful and thorough study. Harrison has argued persuasively that analytical and thorough methods of study being applied to Scripture during the Renaissance and Reformation were reflected in the methodologies being applied to understanding the meaning and purpose of creation.[89] Thus it is important to understand what scriptural methodologies implied about the relation of God to the world and the nature of divine agency in the world. It is useful to briefly review these changes in interpretation from the medieval *quadriga* to early modernity's methods of interpreting Scripture and the world which came to replace it.

---

89. Harrison, *The Bible, Protestantism and the Rise of Natural Science*; Harrison, "The Book of Nature and Early Modern Science."

Quadriga—the Fourfold Method of Interpreting Scripture (and the World)

Scripture interpretation through the Middle Ages used the *quadriga* a refinement of Augustine's earlier fourfold method for interpreting Scripture. This method informed not only faith and ethics but also understanding of the natural world. Harrison proposes that this parallel is important, and that when this method of interpretation changed this resulted in a similar change in the way in which the world was understood. The *quadriga* allowed that truth in Scripture could be gleaned from the text on four different levels:

- the first level being the literal or the surface meaning of the text;
- the second level of meaning lay in allegory, the lesson for belief;
- the third was the moral sense or ethical lesson; and
- the fourth in the ultimate spiritual truth or ideal pointed to by the text.

It was commonplace in the Middle Ages to attribute allegorical and anagogical meanings to the natural order in a manner similar to that used in interpreting Scripture.[90]

Augustine refined the *quadriga* from Origen's earlier three-fold method, which did not include a neo-platonic ultimate truth or ideal.[91] Origen held that some passages were absurd or impossible thereby forcing a spiritual reading in order to find the truth in the text.

> So for that reason divine wisdom took care that certain stumbling-blocks, or interruptions, to the historical meaning should take place, by the introduction into the midst (of the narrative) of certain impossibilities and incongruities; that in this way the very interruption of the narrative might, as by the interposition of a bolt, present an obstacle to the reader, whereby he might refuse to acknowledge the way which conducts to the ordinary meaning; and being thus excluded and debarred from it, we might be recalled to the beginning of another way, in order that, by entering upon a narrow path, and passing to a loftier and more sublime road, he might lay open the immense breadth of divine wisdom. This, however, must not be unnoted by us . . . as the chief object of the Holy Spirit is to preserve the coherence of the spiritual meaning.[92]

---

90. McGrath, *Christian Theology an Introduction*, 176–77.
91. Augustine, *de Doctrina Christiana* 3.30.42–37.56; Origen, *de Principis* 4.1.2.
92. Origen, *de Principis* 4.1.15.

Harrison points out that this was the way Origen could hold that Scripture was both perfect and infallible.[93] Origen's method allowed him to hold that there were different levels of truth in the text and that some texts did not necessarily contain truth at each or every level. For Origen, if a text were not literally true then it must be true on some other level. In this way, Scripture would always be true. Augustine would have disagreed denying any error or even the possibility of error in the literal meaning while still holding that the allegorical, moral and heavenly meanings held the highest forms of truth.[94] However, in order to hold that the text was without error on each of the four levels Augustine had to explain apparent textual errors. As late as the nineteenth century Gaussen supported Augustinian textual infallibility arguing that inconsistencies only seem to be inconsistencies and thus they need to be explained away by inferring transcribal error.[95]

From a contemporary perspective, many allegorical and anagogical meanings seem arbitrary. Nonetheless, in this tradition the methods of deriving these higher truths were bound to earlier traditions of scholarship by respect for their antiquity and the pedigree of a particular illustration. The esteem in which they were held was so high that even absurd and fantastical examples continued to be used. Examples of this type include the phoenix arising from its ashes and the alleged pelican's piercing of its breast to resuscitate its chicks with its own blood; the former illustrating Christ's resurrection and the latter God's resurrection of the faithful.[96] The methods for determining truth in the *quadriga* constituted a system of thought that had wider implications than those applying to Scripture. Harrison's thesis is that the method of interpreting Scripture also paralleled academic interpretation of the natural world.

> Writing in the twelfth century, Hugh of St Victor provides a typical example of the traditional understanding of the metaphor—an understanding which was virtually universal in the Middle Ages, and which persisted right up until the seventeenth-century. All of the elements of the empirical world, says Hugh, are "figures," which have been invested with divinely instituted significance.[97]

---

93. Harrison, *The Bible, Protestantism and the Rise of Natural Science*, 19–20.

94. Augustine, *de Genesi ad Litteram* 8.20.39; Taylor's editorial note in Augustine, *de Genesi ad Litteram*, 2:261n97.

95. Gaussen, *Theopneustia: The Plenary Inspiration of the Holy Scriptures*, 216.

96. Harrison, *The Bible, Protestantism and the Rise of Natural Science*, 22–26.

97. Ibid., 3.

Umberto Eco highlights the contrast between the nature of the fourfold method and post-Enlightenment thought throughout his *Name of the Rose*.[98] In this Eco places an anachronistically modern thinking Holmesean figure, William of Baskerville, into a barely fictional medieval setting involving actual historical figures, situations and disputes. Williams' anachronistic modern logical thinking contrasts bizarrely with the medieval logic of the other characters trying to solve a whodunit. The irony is that, to his chagrin, William solves the mystery and saves his life using the allegorical and anagogical medieval logic he (and the reader in his place) rejects. While post-enlightenment thinkers cannot use these older forms of logic, the *quadriga* as part of the medieval process of thinking provided stability, meaning and explanations about life, the world and their interaction with God.

The *quadriga* led to a particular understanding of creation in relationship to the creator. Examples from nature were spiritually instructive and not all were as erroneous as those of the phoenix and the pelican were. While the inner workings of nature reflected the consistency and immutability of God, they were thought to be knowable only to God. This was even the case if humans could discern a way to describe how nature might work. Brooke recounts the sophisticated expression of this notion in the opinion of one of Galileo's mentors, Cardinal Barberini. Barberini argued

> that definitive conclusions could not be reached in the natural sciences. God in his omnipotence could produce a natural phenomenon in any number of ways and it was therefore presumptuous for any philosopher to claim that he had determined a unique solution.[99]

Unfortunately for Galileo he placed this opinion of his friend Barberini into the mouth of the simpleton in his *Dialogue Concerning Two Chief World Systems*. This was after Bishop Barberini had been promoted and made pope Urban VIII. Galileo had in effect made the pope out to be a fool. His house arrest followed.

It may be concluded that describing the nature of divine agency in the world was peripheral to a world view informed by the *quadriga*. That is happened was taken as a given. However, the manner in which divine agency occurred would have been considered largely unimportant unless it was considered to be of positive spiritual value. This fourfold method of interpretation remained a stable interpretative tool throughout Middle Ages.

---

98. Eco, *The Name of the Rose*.
99. Brooke and Cantor, *Reconstructing Nature*, 110.

The Two Books of God's Revelation

With the Renaissance and Reformation, there came major changes in interpretation of Scripture. Harrison argues that these changes in interpretive methods played a central role in the emergence of natural science in the early modern period.

> It is commonly supposed that when in the early modern period individuals began to look at the world in a different way, they could no longer believe what they read in the Bible. . . . I shall suggest that the reverse is the case: that when in the sixteenth-century people began to read the Bible in a different way, they found themselves forced to jettison traditional conceptions of the world.[100]

Harrison further argues the change in interpretative structure from the *quadriga* to the post-Reformation literalism had a derivative parallel effect on the interpretation of the natural world. With the development of the late medieval metaphor of the two books, the first God's revelation in the written word and the second in nature understood as a book to be interpreted, people were led to a more literal reading of nature in itself as God's handiwork resulting from a more literal reading of Scripture.[101] In the transition between interpretative frameworks, assumptions about the nature of divine action in the world remained both unchanged and unquestioned and were carried from the interpretation of one book to the other. For example, the book of nature's divine authorship made it reasonable to assume, that if correctly interpreted, it was as reliable as the book of Scripture. In the post-Newton period when serious questions were raised by the deists and even Newton himself about possible corruption of the biblical text, "uncorrupted" nature was often seen as being the more reliable "book.'" This new conception of the order of nature was made possible

> by the collapse of the allegorical interpretation of texts, for a denial of the legitimacy of allegory is in essence a denial of the capacity of things to act as signs. The demise of allegory, in turn, was due largely to the efforts of Protestant reformers, who in their search for an unambiguous religious authority, insisted that the book of Scripture be interpreted only in its literal, historical sense.[102]

---

100. Ibid., 4.
101. Ibid., 1–4, 45–47.
102. Harrison, *The Bible, Protestantism and the Rise of Natural Science*, 4.

Harrison highlights important changes. The collapse of allegory meant that nature could not primarily be about signs to eternal meanings. If nature held meaning it would need to be more understood more directly. The place of final authority for determining how Scripture and consequently the world were to be understood had to change. No longer was ancient authority alone enough, irrespective of whether that was that of classical scholar or theologian. It their place protestant scholars explored alternatives including the authority of personal inspiration by God and the application of disciplined reason. The question which then became important was how reliable could these other sources of authority be considered? Indeed, could any authority be considered totally reliable?

Resurgence of Augustinian Studies in the Seventeenth Century

The question of where to locate reliable authority came about as a consequence of the Reformation. It becomes of particular importance to the developing description of divine agency when the notion of the two books was applied to how the world was to be understood and explored. The seventeenth-century resolution to the question of authority led to the affirmation of dispassionate and disciplined reason and—as will be argued—a fresh appropriation of the doctrine of inspiration.

Post-Reformation use of the notion of the two books came at the same time as a change regarding the place of authority. Harrison notes that that medieval use of the metaphor implied that "nature and Scripture must be read together for the meaning of words of Scripture [to be] given by the meanings of natural objects to which they refer."[103] There is, Harrison claims, a transition in the sixteenth and seventeenth centuries whereby "the general tendency now is to elevate nature over some alternative authority."[104] This became relatively straight forward in the case of classical authors; however, this elevation was even argued by some in the seventeenth century to have priority over Scripture when it was thought to be in error by mistake or heretical interpretation. The basis for the later claim, when made, followed Tertullian's reasoning. "We maintain that God must first be known from nature, and afterwards authenticated by instruction: from nature by His works; by instruction, through His revealed announcements."[105] Harrison notes that various commentators in the seventeenth century noted that this argument referred to a time before the canon was settled and applied it again

---

103. Ibid., 194. The original ungrammatically reads "is."
104. Ibid.
105. Tertullian, *adversus Marcion* 2.1.18.

INSPIRATION, PERFECTION, AND GENERIC THEOLOGY

at time while the meaning of the canon was subject to controversy. A typical example Harrison notes is that of Austen in 1653. "Tertullian's priority was again asserted. 'God sent us the *Booke of Nature*, before he sent us the *Booke of Scriptures*.'"[106] As Harrison notes, even where the book of nature is given priority, as by Spratt, Paracelsus, Browne, Hooke and Galileo,[107] the book of nature is clearly understood as written by God and unable to be falsified by human action or error.

Harrison has argued that the transition related to how authority was considered occurred during the seventeenth century's intellectual revival of Augustine's work. In this revival there was a particular emphasis on Augustinian anthropology in relation to the Fall.

> The renewed focus on the Fall and original sin that is characteristic of the early modern period was occasioned by the religious upheavals of the sixteenth-century. These events not only precipitated a crisis of confidence in the traditional sources of knowledge, but also coincided with a revival of an Augustinian anthropology that emphasised the corruption of human nature and the limitations of the intellect.[108]

Harrison traces the seventeenth-century search for certainty in the world by examining different approaches to dealing with implications of Augustine's anthropology of the Fall.[109] These approaches deal with consequences of the fall on anthropology:

1. That error was associated with sin;
2. To what extent did the fall affect the nature of cognitive and physical depredations?
3. What contrasts could be drawn between inherent human fallibility against the reliability of human reason, and;
4. The different ways Augustinian thought developed.[110]

In assessing the question "exactly what physical and cognitive depredations were suffered by the human race as a consequence of Adam's original

---

106. Austen, *The Spiritual Use of an Orchard*, quoted in Harrison, *The Bible, Protestantism and the Rise of Natural Science*, 195.
107. Harrison, *The Bible, Protestantism and the Rise of Natural Science*, 195–97.
108. Ibid., 3.
109. Ibid. 3–8.
110. Ibid.

infraction?" Harrison analyses a range of responses which he summarises as follows: if

> the Fall were understood as having resulted in the triumph of the passions over reason, the restoration of Adamic knowledge would be accomplished through re-establishing control of the passions, thus enabling reason once again to discharge its proper function. If the Fall had dulled Adam's senses, this deficiency might be overcome through the use of artificial instruments capable of restoring to weakened human senses some of their original acuity. If the Fall had altered nature itself, rendering its operations less obvious and less intelligible, intrusive investigative techniques would be required to make manifest what had once been plain. Varying estimates of the severity of the Fall, moreover, gave rise to different assessments of the prospects of a full recovery of Adam's knowledge.[111]

The key question for Newton's contemporaries was how and if it might be possible to seek certainty of knowledge in a fallen world. Harrison argues that these may be distinguished by their emphasis on "the range of available authorities: reason, Scripture, experience, personal inspiration."[112] Foreshadowing the discussion in the next section, divine inspiration came to be thought to be the guarantor of authority and such certainty of knowledge.[113] Harrison details how a variety of methods were used in medicine, alchemy and by a variety of "philosophical enthusiasts" to enable the researcher to discern the shape of unalloyed divine inspiration in nature thereby overcoming their own human fallibility and sinfulness to somehow attain to the knowledge possessed by Adam before the fall.[114] Harrison makes a case that the Reformation and the accompanying resurgence of Augustinian thought led to rethinking issues of authority and reliability of knowledge and epistemology in terms often affected by scholars' theological commitments. For those who were Protestant the source of such authority, tended to come from a Calvinist linking of renewed personal inspiration which confirmed prior divine inspiration in either Scripture or nature.

> [T]hose influenced by the anthropology of Luther and Calvin were to adopt the position of mitigated scepticism characteristic of empiricism and the experimental philosophy. Those who took a more positive view of human nature were more

---

111. Ibid., 6.
112. Ibid., 93.
113. Ibid., 114.
114. Harrison, *The Fall of Man and the Foundations of Science*, 125–30.

inclined to assert the reliability of human reason, the possibility of *a priori* knowledge, and the perfectibility of the sciences. To a degree, then, the methodological prescriptions offered by philosophers in the seventeenth-century mirror their confessional allegiances. Hence, the Catholic Déscartes held fast to a relatively optimistic Thomist account of human nature and aspired to attain, in his own words, a "perfect knowledge of all things that mankind is capable of knowing." By way of contrast, Francis Bacon, raised as he was in a Calvinist environment, thought that knowledge would be accumulated gradually and only with meticulous care.[115]

Bacon, one of Newton's mentors, expresses the meticulous care needed thus:

> [W]e in the beginning of our work pour forth most humble and ardent prayers to God the Father God the Word and God the Spirit that mindful of the cares of man and of his pilgrimage through this life in which we wear out some few and evil days they would vouchsafe through our hands to endow the family of mankind with these new gifts and we moreover humbly pray that human knowledge may not prejudice divine truth and that no incredulity and darkness in regard to the divine mysteries may arise in our minds upon the disclosing of the ways of sense and greater kindling of our natural light but rather that a pure understanding cleared of all fancies and vanity . . . that being freed from the poison knowledge infused into it by the serpent and with the human soul is swollen and puffed up we may neither too profoundly nor immoderately wise but worship truth.[116]

Bacon's shorthand "natural light" only comes as the human student of nature is acted on by God. For Bacon and others any accuracy in natural philosophy was predicated on the inspiration of the researcher by God enabling them to recognize divine inspiration in the particular book of divine revelation they are reading. In nature the researcher might find divine inspiration and hence the light of God's truth.

> First then we admonish mankind to keep their senses within the bounds duty as regards Divine objects . . . For it was not that pure and innocent knowledge of nature by which Adam gave names to things from their properties that was the origin or occasion

---

115. Ibid., 7.
116. Bacon, *Novum Organum*, 10–11.

of the fall but that ambitious and imperious appetite for moral knowledge distinguishing good from evil with the intent that man might revolt from God and govern himself was both the cause and means of temptation ... Lastly we would in general admonish all to consider the true ends of knowledge and not to seek it for the gratification of their minds or for disputation ... For from the desire of power the angels fell and [then] men from that of knowledge.[117]

Bacon made a parallel between Calvinistic secret inspiration confirming prior divine action and that of the inspiration contained in each of the two books. The Lutheran Osiander had earlier expressed a similar belief in his introduction to Copernicus' seminal work on heliocentric planetary motion. "The philosopher will perhaps rather seek the semblance of the truth. But neither of them will understand or state anything certain, unless it has been divinely revealed to him."[118] Harrison has noted that Robert Boyle, Galileo, Déscartes and Malebranche all also place importance on God's inspiration of the investigator to enable the recognition of divine inspiration in nature.[119] Each may have offered differing opinions to whether before the fall Adam naturally possessed perfect knowledge or have gained such knowledge as the result of divine inspiration. In any case inspiration would, it was argued, be necessary for humans to again be able to recognize and understand the truth. Harrison admits Newton does not fit well with his thesis and notes it was, "Newton's rejection of the deity of Christ that indirectly led to his agnosticism about the fallen state of human nature."[120] While many of Newton's contemporaries debated at length about Augustinian anthropology specifically in relation to the fall, Augustinian anthropology has broader application. Newton had little interest in the nature of the fall but, as will be shown, had a great interest in the broader application of Augustinian anthropology particularly in relation to Augustine's description of inspiration thus alleviating Harrison's concern. Inspiration by means of divine agency through an Augustinian anthropology as well as ensuring similar reliability and omnipresent possibility of divine action in nature will be demonstrated to be paramount in Newton's theology. It will be argued that the relationship between these led Augustinian inspiration to be demonstrably more significant for Newton than speculations on the fall.

---

117. Bacon, *Novum Organum* 11
118. Osiander, "Introduction," in Copernicus, *On the Revolutions*, xx.
119. Harrison, *The Fall of Man and the Foundations of Science*, 134–35.
120. Ibid., 239. Harrison focuses on the Fall. Ibid., 233–37.

In the seventeenth-century backdrop to Newton's working life, the continuing Augustinian influence and that of the notion of the two books of God's revelation can be easily demonstrated. Sir Walter Raleigh in his *History of the World*, after praising Ambrose and Augustine, declares that as well as Scripture coming from God's hand there is also: "God's word written on the firmament."[121] Lucius Cary, Lord Falkland wrote and affirmed at length an influential Protestant version of infallibility. "For since it is impossible that Saint John and Saint Peter both inspired by the Holy Ghost which is the Spirit of Truth could teach contradictory doctrines."[122] This infallibility of Scripture being confirmed, albeit imperfectly, by personal inspiration of the Holy Spirit, "for I am confident that those who would know it by the Spirit run themselves into the same Circle between Scripture and Spirit out of which some of your side have but unsuccessfully laboured to get out between Scripture and Church."[123] Thomas Burnet describing his explanation for the creation and eventual end of the world in his *Telluris Theoria Sacra* affirms not merely a biblical literalist point of view but one that depends explicitly on Augustine's explanation in *de Genesi ad Litteram*.[124] This was also known to and publicly approved by Newton. Bishop Stillingfleet who was critical of the implications of Newton's philosophy nonetheless reflected a general feeling of the time in his *Origines Sacrae of a Rational Account of the Grounds of Natural and Revealed Religion*. (1662) "None who heartily believe the Scripture to be the Word of God and that the matter revealed therein are infallibly true, will ever haveth less estimation of it."[125]

The influence of the notion of the two books of God's revelation as a factor influencing the development of an understanding of divine agency can be summarized in a number of points. The notion of the two books of God's revelation operates with assumptions about the perfection of God's actions in the world but also without any reference to Christology. It was suggested on the basis of the analysis of Barth's observations on divine perfections earlier that it is possible to predict that any generic description of divine agency would develop into, in the face of contradictory evidence, an unresolvable choice between finding new ways to assert absolute truths or to abandon traditional ways of speaking of God. Evidence will be presented that this eventually happened with the application of the notion of the two

---

121. Raleigh, *The History of the World*, 1.1.

122. Cary, *Discourse of Infallibility*, 99.

123. Ibid. 55.

124. Burnet, *The Sacred Theory of the Earth*, 1.6.39.

125. Stillingfleet, *Origines Sacrae or a Rational Account of the Grounds of Natural and Revealed Religion*, 9.

books when the priority of the book of nature was advanced over that of Scripture when corruption of the text had been suspected either by error of transmission or by heretical misinterpretation. This issue becomes particularly significant for Newton and will be discussed more fully in the next chapter.

With application of the notion of the two books, God's purposes came to be considered as potentially knowable in themselves. Therefore, rather than being a peripheral topic divine agency became of more central interest. The notion of the two books suggests there may be a common understanding possible between divine agency in nature to that of divine agency in Scripture, but does not lead directly to such an understanding. What has been suggested here is that the need to resolve the question of authority led to consideration of dispassionate and disciplined reason and the reappraisal of the Augustinian doctrine of inspiration.

Inspiration as Guarantee of God's Action—the Third Factor

This section will illustrate how in early modernity the description of perfect divine agency encapsulated within the doctrine of inspiration became an unquestioned assumption which functioned as a precondition to understanding divine action in the world and even the Christian faith. This doctrine of inspiration was manifestly that formulated by Augustine but adapted to early modern use assuming the widely accepted descriptions of divine perfection. Examination of the thought of the important figures of Newton, Darwin and Huxley in following chapters will demonstrate how aspects of this Augustinian inspiration influenced their thought about divine agency in the world.

Inspiration Begins to Bear the Weight of Scriptural Authority

Inspiration did not bear the weight of ensuring the authority of Scripture alone until after the Reformation. Inspiration, as shown in the last section, did not begin to seriously be considered a source of authority for interpreting nature until raised in conjunction with the seventeenth-century application of the notion of the two books of God's revelation. It is this change in the emphasis regarding authority that is important. The sole doctrine of inspiration used in the seventeenth century was that of Augustine's. Whereas Augustine will be shown to have concluded that the divine inspiration of

Scripture was inherently perfect and without error, he did not base the authority of Scripture on inspiration alone. He states clearly, "For my part, I should not believe the gospel except as moved by the authority of the Catholic Church."[126] It is not to be expected that Augustine saw that there was any conflict between either source of authority, as became the case during the Reformation.

However, during the Reformation, Calvin explicitly reinterprets Augustine's comment to assert the authority of Scripture over the church.[127] Calvin further viewed the Bible as given by the hand of God, the writers being the secretaries of the Holy Spirit. His view placed every part of Scripture equal in authority to each other. While his doctrine of inspiration is not extensively developed in *Institutes*, Calvin obviously built his theology strongly on Augustinian ideas.[128] *Institutes* 1.6-10 contains Calvin's development of his doctrine of Scripture, countering Luther's graduation of authority, Rome's insistence on authority and Anabaptist fanaticism. He relies heavily on the claim that Scripture is totally reliable and points to the witness of the Spirit as ensuring correct interpretation.[129] Calvin also simultaneously argues, contestably, that Augustine did not mean the tradition of the Church to hold equal authority.[130] Nonetheless, Calvin while going to great lengths to explain many terms related to how God is to be known throughout the first volume of *Institutes*, never develops the notion of inspiration. This is because the definition of inspiration is not in dispute, unlike the other terms. He does seek to give it higher authority. He knows and assumes that his readers know only one description of inspiration and this description is Augustinian. His presupposition of an Augustinian concept as a foundation to his theology has had profound implications for Protestant theology in the reformed tradition.

Calvin claimed that "the Scriptures are the only records in which God has been pleased to consign his truth to perpetual remembrance, the full authority which they ought to possess with the faithful is not recognized, unless they are believed to have come from heaven, as directly as if God had been heard giving utterance to them."[131] "Error never can be eradicated from the heart of man until the true knowledge of God has been implanted

---

126. Augustine, *contra Epistolam Manicaei Quam Vocant Fundamentum* 5.6.
127. Calvin, *Institutes* 1.7.3.
128. Kraeling, *The Old Testament Since the Reformation*, 24.
129. Calvin, *Institutes* 1.6.1.
130. Ibid., 1.7.
131. Ibid., 1.7.1.

in it."[132] Not only did he argue that the Scriptures were inspired in this direct manner he also referred to the need for the reader of the Scriptures to also be inspired by the same Holy Spirit.[133] Calvin uses inspiration of both the production of Scripture and in the life of faith. He speaks of error-free divine inspiration of the biblical writers, e.g., "When Moses relates the words which Jacob, under Divine inspiration uttered."[134] In contrast secret inspiration is "not as high"[135] and subsidiary in confirming human opinion of that which is perfect. "This is done when by the secret inspiration of his Spirit he displays the efficacy of his word, and raises it to the place of honor which it deserves."[136] Calvin's inspiration follows what will be shown to be an Augustinian graduation reliability in which only scriptural inspiration is of the highest and infallible kind. "God exerts his power, because by his divine inspiration he so breathes divine life into us, that we are no longer acted upon by ourselves, but ruled by his motion and agency, so that everything good in us is the fruit of his grace."[137] Also Calvin's use of inspiration has a broader sense of the Holy Spirit's action in Christian life than that usually attributed to it in later modernity which largely limits it to be part of the doctrine of Scripture.[138] In this sense inspiration stands apart from divine perfection and the notion of the two books and deserves consideration as an additional factor contributing to development of divine agency, albeit a factor closely linked with the others.

> The Scots Confession of 1560 follows and summarises Calvin.[139] This asserts that Scripture is written by the Holy Ghost and that the Spirit affirms its interpretation confirming the plain meaning of the text. The authority of the Scripture is given by God not by the Kirk. The defining authority of the church is formally replaced by Protestants by that of Scripture.[140] It is the agency of the Holy Spirit which ensures this biblical authority.[141]

---

132. Ibid., 1.6.3.

133. Ibid., 1.7.

134. Ibid., 1.8.4. A similar sense is used of divine inspiration in 1.8.7, 1.7.8, 1.13.7, 1.14.7, 4.10.25.

135. Ibid., 1.16.1.

136. Ibid., 3.20.7. A similar sense is used of secret inspiration in 1.18.2, 3.20.5, 3.20.42.

137. Ibid., 3.1.3.

138. Watson, "Hermeneutics and the Doctrine of Scripture," 9n20.

139. Scots Confession of Faith, 18 and 19, in Owen, *Witness of Faith*, 72–73.

140. Ibid.

141. Ibid., 18.

INSPIRATION, PERFECTION, AND GENERIC THEOLOGY   59

Again following Calvin, this Holy Spirit given biblical authority must be brought to individual recognition by ongoing inspiration. "This our faith, and the assurance of the same, proceeds not from flesh and blood, that is to say, from no natural powers within us, but is the inspiration of the Holy Ghost."[142]

In response to these Reformation declarations, the Council of Trent declared in its fourth session in 1546 that, "all biblical books . . . have equal authority, for they are all divinely inspired."[143] However, this authority is qualified by the church being, "the final, because infallible, judge of the true sense of Scripture."[144] There was no dispute over the description of inspiration, only over whether it alone could bear the weight of biblical authority.

A similar lack of dispute about the description of inspiration is evident in between opponents in republican England. Royalist Chillingworth's often reprinted *Religion of the Protestants* held contrary to Trent that divine inspiration alone guaranteed the infallibility of Scripture.[145] But that this inspiration was of a different kind to that in the contemporary church. "Pastors there are still in the church but not such as Titus and Timothy and Apollos and Barnabas not such as can justly pretend to immediate inspiration and illumination of the Holy Ghost"[146] He goes as far as to suggest that divine inspiration may have guaranteed infallibility in describing the natural world.

> I hope you will grant that Hippocrates and Galen and Euclid and Aristotle and Sallust and Caesar and Livy were dead many ages since and yet that we are now preserved from error by them in a great part of physic of geometry of logic of the Roman story. But what if these men had writ by Divine inspiration and writ complete bodies of the sciences they professed and writ them plainly and perspicuously you would then have granted I believe that their works had been sufficient to keep us from error and from dissension in these matters.[147]

Reventlow argues that Chillingworth is "particularly important" in the development of seventeenth-century theology as "he puts the principle that the Bible, and the Bible alone, is the complete rule for the faith and action

---

142. Ibid., 9.

143. Kraeling, *The Old Testament since the Reformation*, 34.

144. Ibid.

145. Chillingworth, *Religion of Protestants a Safe Way to Salvation*. Cary, for example, cites Chillingworth. Cary, *Discourse of Infallibility*, xiii.

146. Chillingworth, *Religion of Protestants a Safe Way to Salvation*, 608.

147. Ibid., 265.

of Protestants."[148] A few years after the royalist Chillingworth's death,[149] the republican Westminster Confession (1647) and then later the Savoy Declaration (1658) explicitly tied Protestant understanding of the authority of Scripture in similar ways to its inspiration by the Holy Spirit.

> II. Under the name of Holy Scripture, or the Word of God written, are now contained all the Books of the Old and New Testament . . . All which are given by inspiration of God, to be the rule of faith and life.
>
> III. The books commonly called Apocrypha, not being of divine inspiration, are no part of the Canon of Scripture; and therefore are of no authority in the Church of God, nor to be any otherwise approved, or made use of, than other human writings.
>
> IV. The authority of the holy Scripture, for which it ought to be believed and obeyed, dependeth not upon the testimony of any man or Church, but wholly upon God (who is truth itself), the Author thereof; and therefore it is to be received, because it is the Word of God.
>
> V. We may be moved and induced by the testimony of the Church to an high and reverent esteem of the holy Scripture; and the heavenliness of the matter, the efficacy of the doctrine, the majesty of the style, the consent of all the parts, the scope of the whole (which is to give all glory to God), the full discovery it makes of the only way of man's salvation, the many other incomparable excellencies, and the entire perfection thereof, are arguments whereby it doth abundantly evidence itself to be the Word of God; yet, notwithstanding, our full persuasion and assurance of the infallible truth and divine authority thereof, is from the inward work of the Holy Spirit, bearing witness by and with the Word in our hearts.[150]

In each example Calvin, Scots confession, Trent, Chillingworth and Westminster only one description of inspiration was in discussion, i.e. that of Augustine. The conclusion that divine inspiration must result in infallibility as a consequence of perfect divine action in the act of inspiration had become the undisputed basis for supporting the protestant understanding of biblical authority. In post-Reformation scholarship inspiration is neither qualified as Augustinian nor *ekstasis* as this is the only way they understood

---

148. Reventlow, *The Authority of the Bible and the Rise of the Modern*, 148.

149. Chillingworth served with Charles I's forces and was captured and died in detention in 1644.

150. Westminster Confession and Savoy Declaration 1, in Owen, *Witness of Faith*, 121–22.

inspiration worked. Similarly when Newton, Darwin and Huxley refer to inspiration they do so without any qualification. Such references can always be understood as deriving from what will be described as the Augustinian *ekstasis* description. These academics and their disputants never critically examine the assumptions of this description of inspiration even while attacking arguments or doubting the reality of inspiration. Irrespective of where divine law is written be it in Scripture or in the book of nature God's act as divine action was understood to be perfect.

What will be explored in the next section is this Augustinian description of inspiration which was recast to support authority following the Reformation. This in turn will be shown to lead to, when combined with established notions of divine perfection and the two books of God's revelation, a particular theological development of understanding of divine agency in the world. It is important to detail what was understood: the assumptions inherent in the description; its development; and its susceptibility to use without recourse to who God is. What is surprising is that uniquely for Augustine and also for Tertullian (on whom Augustine draws heavily) is that this description can be made to work without reference to Christology or the trinity because neither is central to its development.

Augustine's Description of Inspiration by Way of Tertullian

Francis Watson expresses a commonly held view when he described the doctrine of inspiration as a foundation for understanding the doctrine of Scripture.[151] Nonetheless, there exists no comprehensive contemporary study of the doctrine's development, which does not assume that inspiration is primarily about Scripture.[152] This section seeks to challenge that assumption. The aim here is to describe inspiration as understood in the seventeenth century in a way that highlights those aspects of the doctrine would be used to shape understanding of divine agency in conjunction with early modern understanding of divine perfections and the application of the notion of the two books of God's revelation.

151. Watson, "Hermeneutics and the Doctrine of Scripture: Why They Need Each Other," 9n20.

152. The following references each assume inspiration is primarily relates to Scripture. Alonso-Schökel, *The Inspired Word*; Baille, *The Idea of Revelation in Recent Thought*; Benoit, *Revelation and Inspiration*; Gaussen, *Thoepneustia: The Plenary Inspiration of the Holy Scriptures*; Marshall, *Biblical Inspiration*; Sanday, *Inspiration*; Warfield, *The Inspiration and Authority of the Bible*; Barr, *Fundamentalism*; Barr, *Escaping from Fundamentalism*; Tillich, *Systematic Theology*, 1:124–26.

It will be argued that the focus of inspiration on Scripture is a later development of modernity that actually is as problematic as Francis Watson indicates. Watson is not alone in expressing dissatisfaction with the Western or Augustinian description of inspiration. Sasse expressed doubt about the adequacy of the Western doctrine of inspiration when he referred to the doctrine of inspirational infallibility as being "pagan." He, however, neither elucidated this statement nor attempted to explain it, instead focusing on Reformation developments.[153] Sasse implies that the close linkage between inspiration and Scripture is a post Reformation development in doctrine. This section will show that Sasse was correct in his assertion regarding the development of the Western description of inspiration. This Western description will be shown to be essentially Augustinian and arose from the description assuming elements of Greek anthropology and philosophy, lightly modified by the Christian theologians Tertullian and Augustine, then used in a consistent shape through the Reformation and beyond.

This section explores the development of the doctrine of inspiration and will show that the focus of inspiration moves to Scripture as its highest expression only with Augustine's refinement of the doctrine. The last section showed how inspiration did not bear the burden of authority of Scripture before Reformation. What is new to identify that inspiration was considered to have a broad application in the early church and then to show how this narrows to the special case of Scripture in early modernity. To assume that that inspiration is mainly a foundation to the doctrine of Scripture misses the historical reality of the doctrine's use. Ultimately it may be posited, as Watson suggests, inspiration is unnecessary as a foundation to the doctrine of Scripture.

There are aspects of Augustine's *ekstasis* description of inspiration that will be significant later in discussing the development of Science. These include the notion, according Augustine's description that the Holy Spirit's inspiration is understood to be solely God's action and is thus automatically perfect. It also includes how this description tied Christian anthropology and Pneumatology to a particular form of description of the relationship between the soul and body and the place of the mind and senses within them. Because theologians and natural philosophers considered no other way to describe the Holy Spirit's inspiring work this description influenced the development of science, divine agency and later played a part in causing problems for the ongoing dialogue between theology and science when both the perfection of divine action in nature and metaphysics came to be questioned.

153. Sasse, "The Rise of the Dogma of Holy Scripture," 44–54.

INSPIRATION, PERFECTION, AND GENERIC THEOLOGY

This chapter argues that the development of the dominant Western doctrine of inspiration has been more influenced by certain Greek psychological assumptions about the nature of *ekstasis* than by biblical exegesis or theology. The transliterated *ekstasis* will be used in preference to the English "ecstasy" as the key idea will be shown to not have the connotation of being mystically or rapturously carried away which the English word has, but rather the Greek word's sense of standing beside or "out of." A case in point is the English translation of Tertullian's *de Anima* 9.4 "whilst this sister of ours was rapt in the Spirit." The *Ante-Nicene Fathers* edition includes the extra word "rapt." Rapt is a typical translational interpolation conflation of terminology with the English word ecstasy. This exemplifies why the term *ekstasis* is used here in preference to the English ecstasy. The Latin is simply "whilst our sister was in the Spirit." Latin: *cum ea soror in spiritu esset*. Similarly there is *The New English Bible's* spurious translation of γλῶσσαι (glossai = tongues) as the "language of ecstasy."[154]

"Secular" Greek assumptions will be shown to have been drawn into theological description through Tertullian's engagement with second-century philosophy and medicine. This is later revised by Augustine resulting in a description which became a keystone of Western theology for more than a millennium. This description is not christological and is demonstrably largely developed without reference to who God has revealed God to be. While Augustine includes his notion of the trinity being reflected in human body, soul and spirit and that this is important to his understanding of error, this reflection of the trinity was not present in Tertullian's original work. Whilst Augustine's description became the only conceivable description of inspiration for many theologians and natural philosophers, it is Augustine's debate with Jerome that suggests that the Augustine's description of inspiration was not the only understanding possible or held historically. This opens the possibility of suggesting a description of inspiration that does not require either perfect divine action or a particular understanding of soul and body in theological anthropology.

Beginnings of the Doctrine of Inspiration

The earliest descriptions of the Holy Spirit's inspiring work are in Scripture. More detailed descriptions arose in the process of reflection upon Scripture and tradition, and in the debates of the early Church. However, only two references to inspiration occur within Scripture. They are—"Above all, you

---

154. Tertullian, *de Anima* 9.4; Waszink, *Tertullian's de Anima*, 9.4; 1 Cor 12–14 (NEB).

must understand that no prophecy of Scripture came about by the prophet's own interpretation. For prophecy never had its origin in the will of man, but men spoke from God as they were carried along by the Holy Spirit";[155] and—"All Scripture is God-breathed (Θεοπνευστια—*theopneusia*) and is useful for teaching, rebuking, correcting and training in righteousness."[156] Hence, it is possible to infer that God's moving of people by the Spirit equates with God breathing into humans by the Spirit, hence the term "inspiration." It is also present in the Hebrew text. Childs claims that refinement came in understanding the Holy Spirit's paracletic work as part of God's progressive self-revelation to God's people throughout the Hebrew Scriptures, in spite of the absence of inspiration language.[157] It is God who inspires, who gives breathe in Job 27:3.

Humans are utterly dependent on God for each breath as an ongoing gift of God's Spirit. This notion of inspiration is rudimentary in that it is noted that God moves people, but how God does this, is not explained. In the earliest years of the Church, it was not necessary to know how God carried along and inspired, merely to acknowledge that God did. While Timothy specifically links inspiration with Scripture, Peter infers the Holy Spirit's inspiring work recorded in scriptural prophecy is of the same class as all other Holy Spirit-inspired prophecy. Indeed, it is possible to speak of the Holy Spirit's inspiring work more broadly. Descriptions of inspiration came to include ideas such as the Spirit being the cause of God's self-communication with human beings, the one who carries the Christian along, or who is the breath in their nostrils, or the one who inspires them to life in Christ. This idea is not unique to the Scriptures or to Christianity and is used by Philo and Josephus, as well as in early rabbinic thought.[158]

The biblical beginnings of understanding inspiration, while rich in imagery, left open the question of the nature of divine agency regarding how the Holy Spirit works within human beings. The more common description used in the literature is that of "mantic" possession[159] in which inspiration is considered as a species of possession in which the spirit's presence takes over the consciousness of the person.[160] To this broad group Kelly includes Athenagoras, the Montanists, Tertullian, Chrysostom and Ambrose. Socio-historical studies agree with Kelly's assessment of the breadth of adherence

---

155. 2 Pet 1:20–21 (NIV).
156. 2 Tim 3:16.
157. Childs, *Old Testament Theology in a Canonical Context*, 28–49
158. Aune, *Prophecy in Early Christianity and the Ancient Mediterranean*, 103–52.
159. Kelly, *Early Christian Doctrines*, 61–64.
160. Ibid.

to the "mantic" possession theory.¹⁶¹ Even though in the patristic period variations on the "mantic" possession theory were commonplace, it does not adequately describe the full range of human activities in which inspiration was understood to occur. Ultimately, it is an inadequate description of inspiration. Aune notes that cases are not always or indeed not often associated with mania or a period of frenzy. Inspiration, he concludes, depends "not on form but supernatural origin."¹⁶² As will be demonstrated in the following discussion of Tertullian, "mantic" possession does not also sit well with then contemporary Greek philosophical or medical explanations of consciousness, sleep, death and their relationship to divine action in a person as understood and revised by early theologians. Despite "mantic" possession's inadequacy as a describing theory of inspiration, it has been regularly used.¹⁶³ Nevertheless, what is clear in early patristic usage is that the question of inspiration is primarily deals with the problem of how the Holy Spirit is involved in God's self-communication to humans. This understanding develops before and continues after the development of the doctrine of Scripture or establishment of the New Testament canon during later third of the second century.¹⁶⁴

Tertullian

Tertullian expressed the next major refinement of inspiration. Augustine will be demonstrated to have drawn heavily Tertullian's description in *de Anima*. This description is an *ekstasis* description of inspiration which in turn revises Aristotelian metaphysical anatomy and incorporates Tertullian's own revision of Greek medical ideas about the nature of the soul. Tertullian's theology has long been suspect because his later work was thought to have been influenced by heretical Montanist ideas including the scornful opinion of Augustine expressed in *de Bono Viduitatis* and in *de Anima et Eius Origine*.¹⁶⁵ This has set the standard for the assessment of Tertullian's

---

161. Aune, *Prophecy in Early Christianity and the Ancient Mediterranean World*; Grabbe, *Priest, Prophets, Diviners, Sages: A Socio-Historical Study of Religious Specialists in Ancient Israel*.

162. Ibid., 337.

163. There have been many other surveys, a selection includes. Gaussen, *Theopneustia*, 24–57; Pannenberg, *Revelation as History*, 26–34; Sanday, *Inspiration*, 1–122; Torrance, *The Mediation of Christ*.

164. Frend suggests AD 170–200. Frend, *The Rise of Christianity*, 250–51.

165. Augustine, *de Bono Viduitatis* 6, 8; Augustine, *de Anima Et Eius Origine* 2.9.

later works.[166] However, recent reconsideration by Barnes, Rankin, Kelly and Osborn challenges this view.[167] In contrast to a supposed heterodoxy, Tertullian's development of a description of inspiration will be shown to be a careful attempt to refine an orthodox appreciation of the Holy Spirit's work. Tertullian's *de Anima* was one of his later works and appears to give a comprehensive summary of his views on the *ekstasis* nature of inspiration. Waszink concludes that *de Anima* was written after the now lost *de Ecstasi*. *De Anima* appears to contain a summary of this earlier 7-volume work. We cannot recover the detail of *de Ecstasi*, but Tertullian's critical use of Aristotelian categories of description in *de Anima* does set broad limits to how he may have developed his thought. The loss of his *de Ecstasi* limits retrievable detail regarding his views on inspiration.[168] *De Anima* demonstrably became foundational for Augustine's further revision. It is unusual among the extant works of Tertullian in positively referring to and actively engaging with a broad range of philosophical and "scientific" literature. "Science" is to be understood in this context as the Greek *scientia*, knowledge, which bears limited resemblance to the modern term. Tertullian builds a case for the nature and properties of the soul and the Holy Spirit's inspiring work using these ideas as well as Scripture. While later theologians rejected his high view of the authority of contemporary or continuing works of the Holy Spirit, his description of inspiration continued to be influential as it was reshaped by Augustine.

Tertullian and Philosophy

Upon re-examination of Tertullian's thought, a growing group of scholars place him in the theological and intellectual mainstream of his day. McDonnell concludes, "There is little doubt of Tertullian's perspicuity in matters of philosophy."[169] His comprehensive use of ideas, illustrations and quotations leads Fredouille to conclude that Tertullian had an excellent grasp of his culture, rhetoric, philosophy and faith.[170] In other words, despite the burden of later orthodox suspicion, Tertullian offers significant critical appraisal of

---

166. E.g., Nock, "Tertullian and the *Ahori*," 129–41; Waszink, *Tertullian's de Anima*, Passim; Burleigh, "The Doctrine of the Holy Spirit in the Latin Fathers," 113–32.

167. Rankin, *Tertullian and the Church*, 37; Barnes, *Tertullian: A Historical and Literary Study*, 3–12, 235–41; Osborn, *Tertullian*, 143.

168. Waszink, *Tertullian's de Anima*, 5*.

169. McDonnell, "Communion Ecclesiology and Baptism in the Spirit: Tertullian and the Early Church," 671. Osborn, *Tertullian*, 31; Barnes, *Tertullian*, 206, 123.

170. Fredouille, *Tertullien et la Conversion de la Culture Antique*, 481–85.

the philosophy of his time. Tertullian develops his description of inspiration with philosophical sophistication and careful theological analysis. The intellectual depth of his theological engagement with and re-evaluation of Aristotelian metaphysical notions of the soul and Greek medical knowledge set a pattern also used by Augustine.

Tertullian's astute grasp of philosophy differs to the impression gained at first glance from Tertullian's well-known statement. "Indeed heresies are themselves instigated by philosophy";[171] and his famous polemical rhetorical question, "What has Athens to do with Jerusalem?"[172] The implied answer to the question seems to be, "nothing!" However, on closer inspection, a more accurate answer overall in his writings would be, "nothing to do with the foundation of faith, however if the user avoids heresy philosophy might be used very cautiously in decorating its theological architecture." Tertullian's suspicion of Greek philosophy had a particular structure that is important in the development of the argument in *de Anima*. Tertullian utilises Aristotle throughout *de Anima* but heavily revises his ideas in the light of medical advances of his day and by reference to Scripture.[173] Indeed, Tertullian's representation of Aristotle's ideas was in keeping with contemporary academic best practice, quoting Pliny, Lucretuis, Seneca, Suetonius and Sallust; using Latin rhetorical conventions, classical metrical devices. While Tertullian uses platonic and stoic terminology, he harbours extreme suspicion about the nature of the source of much of this philosophy. Tertullian here refers to Plato's Athenian academies, Stoic porches and Socrates' prison. In contrast, Tertullian appears to exempt the Lyceum from generalized dismissal in *de Anima*.[174] Nor is Tertullian optimistic about philosophy's ability to either find truth or sustain the spirit.[175] In contrast, Tertullian appears to exempt the Lyceum from generalized dismissal in *de Anima*.

While largely critical of the "poison of the philosophers," Tertullian does allow that on occasion it "sometimes happens even in a storm, when the boundaries of sky and sea are lost in confusion that some harbor is stumbled on . . . by some happy chance."[176] On the other hand, Tertullian has a high regard for medical thought, as shown in his support of Soranus whilst dismissing Plato and Socrates. "Soranus, who is a most accomplished

---

171. Tertullian, *de Praescriptione* 7.

172. Ibid.

173. Tertullian, *de Anima* 6, 8, 25–26.

174. Waszink, *Tertullian's de Anima*, 35n; Tertullian, *de Anima* 8.4, 5.6, 20.1, 24.5, 42.2, and 44.

175. Tertullian, *de Anima* 6.

176. Ibid., 2, 10.

authority in medical science,"¹⁷⁷ provides factually that "the soul is nourished by corporeal aliments, let the philosopher (adopt a similar mode of proof, and) show that it is sustained by an incorporeal food."¹⁷⁸

By any current standard, Tertullian's ideas about the soul are unusual. However, knowledge now commonly understood about human anatomy and physics was unimaginable in the early third century. Indeed, the functioning of nerves and the brain only began to be described in the nineteenth century. Mindful of this observation, Tertullian's *de Anima* nevertheless constitutes a state-of-the-art contribution to intellectual debate in his own time. While Tertullian's synthesis of these ideas is unsatisfying for contemporary theology and science, his description of the soul adequately fits into and expands the system of his day within the limits of theological and medical verification open to him.

Tertullian carefully used and revised Aristotle's explanation of the soul. It is useful to describe briefly how Aristotle described the anatomy of soul, mind and flesh, before describing why Tertullian revises it. In Aristotle's terminology, it is the combination of the substances of soul and body, which constitute an individual. Aristotle argues that the soul is a substance with matter and form. The element that knows and thinks the *nous* or mind is in the soul but not in the body. The mind within the soul is used in sensation, which combines the ability to sense with some organ of sensation.¹⁷⁹ This combined action gives rise to thought. Sleep is the seizure of the primary sense organ rendering it unable to actualize its powers. The mind or the seat of identity is not an object or substance but a property of the soul.¹⁸⁰ The mind is primarily metaphysical; raising the possibility that reason may enter from some external "divine" source.¹⁸¹

The Soul in de Anima

Scripture says very little about the structure of the human soul, its psychology or anatomy or the mechanism for its interaction with God. Typically, the narratives leave us with very bare statements such as the "word of the LORD came to Jeremiah."¹⁸² Alternatively, the narrative describes the circumstances of God's interaction with people but usually not the inner

177. Ibid.. 6.
178. Ibid.
179. Spicer, *Aristotle's Conception of the Soul*, 3–7, 29–34, 60–63, 86–93, 95–96.
180. Ibid., 3–7, 29–34, 60–63, 86–93, 95–96.
181. Spicer, *Aristotle's Conception of the Soul*, 103–5.
182. Jer 1:2.

workings of their soul or mind. Significantly, Tertullian seeks to explain the "how" of this interaction by means of theological reflection in *de Anima* of the examples of Adam, Nebuchadnezzar and Daniel.[183] Tertullian develops the nature and functions of the soul,[184] the nature of error[185] and the origin and development of the soul and what happens to the soul in sleep[186] and after death. The latter half of the book contains the bulk of his revision of Aristotle's ideas and his rebuttal of other theories based on Aristotle revising them with medical and biblical evidence.

Tertullian revises Aristotle based on his assertion that true understanding of humanity is only possible with the revelation of God and the guidance of the Holy Spirit.[187] He uses the evidence of Scripture and Christian life to amend earlier attempts by Greek thinkers to understand the natural world.[188].

Whereas, Aristotle denies the soul is in motion but is rather its cause, Tertullian argues that the soul is always in motion. Nevertheless, he describes the body and soul as two pure substances each with characteristically different accidents or properties. The body has weight, height, color, and texture. The soul, gives motion, has sensation, thought, and emotion and enables nutrition and procreation. Tertullian's integration of functions of the soul (its sensorium) is similar to Aristotle's. Sensation is only possible when soul and body are together. This follows and argues his position against other theories. "Again, whence arises sensation if not from the soul? For if the soul had no body, it would have no sensation. Accordingly, sensation comes from the soul, and opinion from sensation; and the whole (process) is the soul."[189] While Tertullian, like Aristotle, regarded the mind as an attribute of the soul's set of interrelated and dependent functions, Tertullian distances himself from Aristotle's idea that there is an impassable divine element of the mind.[190]

> [T]o acquire knowledge is to exercise the senses; and to undergo emotion is to exercise the senses; and the whole of this is a state of suffering. But we see that the soul experiences nothing of these things, in such a manner as that the mind also is affected

183. Tertullian, *de Anima* 11, 45.
184. Ibid., 1–24.
185. Ibid., 25.
186. Ibid., 12. .
187. Ibid., 1.
188. Ibid., 43.
189. Ibid., 17.
190. Ibid., 12, 22.

by the emotion, by which, indeed, and with which, all is effected. It follows, therefore, that the mind is capable of admixture, in opposition to Anaxagoras; and passable or susceptible of emotion, contrary to the opinion of Aristotle.[191]

The mind is an attribute of the soul which includes the intellectual powers and the sensuous faculties functioning as an integrated whole.

Aristotle appears not to make any clear reference to *ekstasis*. Tertullian resolves this omission in Aristotle by reference to his other sources. Tertullian's description of *ekstasis* and the soul in chapters 9 and 10 includes what is for him an important case study. Tertullian puts particular emphasis on one discussion held with this sister. "We discussed thoroughly what I did not know about the soul, while our sister was in the Spirit."[192] Tertullian expresses the care with which his group examined this discussion to justify that this case provided evidence that the soul or spirit had corporeal attributes of colour, texture and spatial extension.

> After the people are dismissed at the conclusion of the sacred services, she is in the regular habit of reporting to us whatever things she may have seen in vision (for all her communications are examined with the most scrupulous care, in order that their truth may be probed). "Amongst other things," says she, "there has been shown to me a soul in bodily shape, and a spirit has been in the habit of appearing to me; not, however, a void and empty illusion, but such as would offer itself to be even grasped by the hand, soft and transparent and of an ethereal colour, and in form resembling that of a human being in every respect." This was her vision, and for her witness there was God.[193]

Tertullian uses these attributes in building his case that the soul is to be described in revised Aristotelian terms as the spiritual, though corporeal, element of human anatomy. Irrespective of what credence may be attributable to these visions, other explanations are possible. The attribute of spatial extension, for example, meets a precondition of the soul's existence as deduced by the agnostic Thomas Huxley based on biological experiment in the nineteenth century.[194] This same bodily shape could also be explained as the Spirit continuing to be shaped by the preservation of the ascended Christ's humanity.

---

191. Ibid., 9.
192. Waszink, *Tertullian's de Anima*, 9.4.
193. Ibid.
194. See discussion in chapter 4.3 of Huxley, "Has a Frog a Soul?"

INSPIRATION, PERFECTION, AND GENERIC THEOLOGY 71

The manner of the Holy Spirit's agency in inspiration must of necessity require *ekstasis* in Tertullian's formulation of the soul as it interrupts the integration of the sensuous faculties. The Holy Spirit must stand into the place of the human mind in the soul in order to use the human sensorium. The human consciousness must stand beside its normal thought processes. Prophecy, which Tertullian attributes to Adam and others as examples, must arise by God's standing aside of the person while the Holy Spirit operated that particular gift.[195] In this infusion the normal integration of the parts of the soul is changed, hence the mind cannot think as it does normally. Thus for Tertullian, following this construction always means that *ekstasis* occurs in any prophet. The example of Nebuchadnezzar is used by Tertullian to demonstrate how it is that the same *ekstasis* occurs even in situations where it might not be apparent that it had occurred.[196] By indicating the scope of how inspiration occurs and by including nonbelievers as examples, Tertullian indicates that *ekstasis* must come by God's action and choice. Tertullian further applied *ekstasis* inspiration to a broad range of activities. "Whether it be in the reading of Scriptures, or in the chanting of psalms, or in the preaching of sermons, or in the offering up of prayers, in all these religious services matter and opportunity are afforded to her of seeing visions."[197] Tertullian does not expound how all inspired activity involves *ekstasis*, but he does assume it throughout *de Anima*. Although the cases described are unusual for contemporary readers, *ekstasis* is essential to Tertullian's description of the divine agency of the Holy Spirit. Also, Tertullian adds a new idea—Sleep must also involve an interruption of these functions and thus is inaugurated by a similar *ekstasis*. "This power we call *ekstasis*, in which the sensuous soul stands out of itself, in a way which even resembles madness. Thus in the very beginning sleep was inaugurated by *ekstasis*: 'And God sent an *ekstasis* upon Adam, and he slept.'"[198] Dreams then can be an example of the function of memory apart from their normal control.[199] Communicative dreams and prophecy come as the Holy Spirit works through a "normal" nightly state of *ekstasis*. The final chapters deal with death as a special case of ultimate *ekstasis* prior to the resurrection of the body foreshadowed in life by sleep.[200]

195. Tertullian, *de Anima* 11, 21.

196. Ibid. 47.

197. Waszink, *Tertullian's de Anima*, 9.4 .

198. Tertullian, *de Anima* 45. Lit., *Hanc uim ecstasin dicimus, excessum sensus et amentiae instar. Sic et in primordio somnus cum ecstasi dedicatus: et misit deus ecstasin in Adam et dormiit*. Waszink, *Tertullian's de Anima*. 45.

199. Tertullian, *de Anima* 45.

200. Ibid., 50–58.

Tertullian relies on Soranus[201] to conclude that the soul is formed at the same time as the body in the process of conception[202] and on Cleanthes that the soul comes from the sperm.[203] Tertullian shows familiarity with the technical detail of Soranus' *Gynaecology* by summarizing a significant section of the treatment for breach births. He cites the prophetic prenatal ability of Rebecca's twins foreshadowing the future struggles between the nations they were to represent and that of the unborn John the Baptist at the news of Mary's pregnancy.[204] The ability of the foetus to experience *ekstasis* proves the soul is present before birth.[205] However, these biblical examples only partly argue this case rather than settling human opinion as Tertullian claims. While supporting the presence of the soul prior to birth they do not support the generation of the soul contrary to Plato's (and Gnostic) claims of the soul's eternal existence. In Chapter 27 Tertullian's argument mixes both Genesis 1 and physiological function to argue for the conception of the soul. The Scriptural injunction to sexual fidelity he interprets as being a limit against harmful excess. This is not argued by the biblical text. It is an interpretation which agreed with contemporary medical opinion. Strangely, for Tertullian, he here uses biblical examples to support a number of conclusions gained from medicine, rather than vice versa.

Because it is the Holy Spirit's divine agency in the *ekstasis* state, this leads logically to Tertullian's discussion of error. In Chapter 24 Tertullian describes error in terms of forgetfulness drawing on his Aristotelian analysis in Chapter 12.[206] Tertullian doubts Plato's opinion that the soul is divine asking how such "divine a faculty as the soul was capable of losing memory."[207] Tertullian argues that while humans are prone to the error of forgetfulness, God is not. Consequently error has no function in the *ekstasis* state when inspiration occurs. *Ekstasis* means that normal self-awareness is stood to one side when God moves. Hence, inspired action is directly God's action in the soul with the body. As it is God's action, this action can only be perfect. Error has no place in this explanation of *ekstasis* revelation. If it occurs, then this is in the admixture of normal thought and emotions following the *ekstasis* in interpretation and application of the revelation. Error results

---

201. Ibid., 25–27; Ibid. 25; Soranus, "Gynaecology," 4.9.61–4.13.70.

202. Ibid., 27.

203. Osborn, *Tertullian, First Theologian of the West*, 1997, 167; Tertullian, *de Anima* 25, 27; Soranus, *Gynaecology* 1.12.43–44.

204. Tertullian, *de Anima* 26.

205. Ibid. 26. Soranus, *Gynaecology* 1.7.30.

206. Spicer, *Aristotle's Conception of the Soul*, 95–96. Spicer incorrectly refers to chapter 25.

207. Tertullian, *de Anima* 24.

INSPIRATION, PERFECTION, AND GENERIC THEOLOGY 73

from the work of "that manifold pest of the mind of man, that artificer of all error."[208] Tertullian envisages the mind as an incorporeal attribute of the corporeal soul.

Tertullian's revised Aristotelian metaphysical anatomy of the soul implies that any knowledge-based activity during inspiration, such as prophecy, must displace normal thought processes. To work they both need access to the same sensory apparatus within the soul, understanding, memory, hearing and the ability to speak or write. The Holy Spirit cannot be consciously acting as a director to these elements of the sensorium at the same time as the individual. Tertullian describes inspiration as God engrafting a new or additional property onto the base nature of the soul, "A corrupt tree will never yield good fruit, unless the better nature be grafted into it; nor will a good tree produce evil fruit, except by the same process of cultivation."[209] This metaphysical anatomy suggests revelation of necessity must always involve *ekstasis* of self-awareness. The branches and leaves of the base plant are moved to one side.

Summary and Implications

Tertullian's critical synthesis of Christian revelation and *scientia* in his description of the soul was persuasive within the confines of scholarly verification of his day and throughout most of subsequent intellectual history up until the nineteenth century. It is only with the discovery of electric current and modern dissection that it became possible to find a role for nerves other than growing nails at their ends and for the brain other than as hair fertilizer.[210] It is now known that the seat of the intellect is the brain, which is a part of the substance of the body, though how the soul relates to this remains a mystery. Indeed, whether the metaphysical soul actually exists is questioned by recent research which indicates that all functions traditionally attributed to the soul can be identified in biochemical reactions in the brain.[211] Nevertheless, Tertullian marries the description of inspiration to a metaphysical soul thereby making his theology depend on the assumptions of what has become an obsolete understanding of the natural world.

In summary Tertullian's description of inspiration begins with and then revises an Aristotelian anthropology. In this the soul is understood to be an element of a human which is of necessity stood aside during the direct

208. Ibid., 57.
209. Ibid., 21.
210. Tertullian, *de Anima* 51.
211. Russell et al., eds., *Neuroscience and the Person*.

action of the Holy Spirit. Therefore *ekstasis* is automatic when the Holy Spirit acts. Activity during inspiration is solely the Holy Spirit's action and thus as God's action alone it is without error. Error creeps in only during recollection of events occurring whilst inspired. As there is in Tertullian's description no place for error in Holy Spirit-inspired activity, it follows according to his logic that contemporary prophetic messages were as authoritative as those that were historical. Therefore, assuming there was no mixture of human conjecture and opinion, contemporary messages inspired by the Holy Spirit could be considered as authentic and authoritative as the old. This suggests a theological basis for his support of the New Prophecy movement. Tertullian's very high view of Holy Spirit-inspired activity was consistent, without distinction, for all forms of divine agency by the Holy Spirit.

Does his description stand theological testing? While noting Tertullian was far removed from the theological refinements of fifth-century christological debates, a problem ensues if his anthropology of the soul and *ekstasis* is applied to the humanity of Christ. Tertullian did draw parallels, describing the soul of Christ as similar to and the perfection of the human soul.[212] If Tertullian's ideas are brought together, then logically Jesus while revealing God must have been in *ekstasis*. Therefore the will or mind of God would have displaced his human will or mind. Such application of his understanding of the Holy Spirit's paracletic work to Christ's humanity tends either to monophysitism or to monothelitism. While being a clear logical inference in his theory, there is no evidence that this was a concern for Tertullian or even a step he may have made. This kind of concern relates more to the sixth century, not the second. This suggests that Tertullian's description of the Holy Spirit's action in the human person is neither coherent with later Nicene orthodoxy nor in retrospect, scientifically valid. This paucity of Christology in Tertullian's description of inspiration left it vulnerable to being recast non-christologically and generically in the seventeenth century. It will be shown that Augustine extensively uses and revises Tertullian's *de Anima*. Part of Augustine's revision involves the idea that complete *ekstasis* somehow transcends human limitation, error and forgetfulness to enable perfect divine self-communication particularly in the case of Scripture. However, Augustine did not revise Tertullian's sparse Christology, consequently leaving Augustine's description also open to being recast in terms of non-christological perfect-being theologies.

---

212. Tertullian, *de Anima* 16.

Augustine

Tertullian's understanding of divine agency in humans by the Holy Spirit draws no distinction in kind between any God-inspired human responses, whilst assuming a particular relationship of the soul to the body. He specifically drew no distinction in kind between the Holy Spirit's action in inspiring the biblical text or contemporary prophecy or, by extension, inspired reading, teaching or preaching. This distinction developed in the further revision of inspiration in the west by Augustine. Augustine did not merely require simple and total *ekstasis* of an assumed metaphysical soul in the under the divine agency of the Holy Spirit. For him there are degrees of *ekstasis* of which inspiration's highest expression is exemplified by the inspiring of Scripture. For Augustine in the case of Scripture, the difference in degree of inspiration makes it in effect essentially of a different kind, making it wholly divine; inexorably implying such inspiration is perfect divine action. Augustine's longest and most detailed analysis of the soul and inspiration is included in his *de Genesi ad Litteram*.[213] There is some additional material in his *de Anima et Eius Origine*.[214]

This part of the discussion has four goals. The first is to show how Augustine linked inspiration to scriptural perfection and that this dovetails neatly with seventeenth-century understanding of divine perfections. The second is to demonstrate that which has not been previously documented, that Augustine revises Tertullian. In particular, Augustine's *de Genesi ad Litteram* follows a similar structure of argument to Tertullian's *de Anima*, in which Augustine only refers to Tertullian by name when he disagrees with him. Thirdly, Augustine makes the notion of a metaphysical soul tied to the sensorium of the body an essential part of theological anthropology and the nature of divine agency within it. Fourthly, in spite of the later dominance of Augustine's description of inspiration, his dispute with Jerome over biblical translation suggests that it may be possible to pose an alternative to *ekstasis* inspiration.

Augustine's extremely high view of Scripture will be argued results from his refinement of Tertullian's description of inspiration. Augustine concludes that it is only truly inspired writing cannot contain falsehood. If it were false in any part then the whole of Scripture would be liable to falsification. Augustine's absolutism on this matter, marrying inspiration to perfect divine agency, sets a pattern of debate that has been often repeated.

---

213. Augustine, *de Genesi ad Litteram. De Genesi ad Litteram libri duodecim* (*Literal Meaning of Genesis in Twelve Books*) should not to be confused with *de Genesi ad Litteram imperfectus liber duodecim* (*The Literal Meaning of Genesis: An Unfinished Book*).

214. Augustine, *de Anima et Eius Origine*.

Following chapters will show how this combined with the early modern understanding of divine perfection and the notion of God's two books of revelation helped natural philosophers to formulate their understanding of divine agency in nature in the time of Newton. This came to be understood as an "all-or-nothing" reliability of such revelation, and shaped both opposition to Darwin's theories as well as Darwin's own agnosticism and is also implicit in Huxley's rejection of traditional religion.

Initial observation suggests Augustine was not favourably disposed to Tertullian. Augustine's disparaging comment, "the ravings of Tertullian"[215] is well known. Augustine seems to follow Jerome's brief biography of Tertullian which labels him a schismatic and therefore heretical.[216] Augustine also explicitly criticises Tertullian's humour[217] and biblical interpretation.[218] On the other hand, Augustine's use of medical examples, biblical references and lines of reasoning parallel those of Tertullian. There is a particular parallel between Tertullian's *de Anima* and Augustine's *de Genesi ad Litteram*. *De Genesi Ad Litteram* is Augustine's exhaustive literal commentary on the creation narrative in Genesis 1-3. In this work, Augustine engaged with the best of contemporary philosophy of nature and medicine.[219] *De Genesi Ad Litteram* consequently contains a detailed description of how God interacts with the human person in particular the mechanism for Holy Spirit's action in the human soul.

Augustine, like Tertullian, utilises and trusts references to contemporary medical terminology, albeit with a further two centuries of scholarship behind him. Taylor has noted references to, and unnamed quotations from, later medical writers: Celsus, Scribonius Longus, Pliny Priscianus, Vindicianus, Caelius Aurelianus and Galen.[220] Augustine's medical refinement is notable in that he seats the soul in the brain[221] differing from Tertullian who places it around the heart. This could reflect both medical developments in the intervening centuries and Augustine's ease with Plato. While knowledge of the brain had become more medically advanced than in Tertullian's time this understanding is also long obsolete. Tertullian had specifically rejected Plato's suggestion that the head is the seat of the soul.[222] Augustine describes

---

215. Ibid., 2.9.
216. Barnes, *Tertullian: A Historical and Literary Study*, 4-12, 235-40.
217. Augustine, *de Civatae Dei* 7.1.
218. Augustine, *de Bono Viduitatis* 6.
219. It is similar to his grasp of culture in *de Civatae Dei*. Augustine, *de Civatae Dei*.
220. Taylor, editorial note, in Augustine, *de Genesi ad Litteram*, 2:247n32, 295n93.
221. Augustine, *de Genesi ad Litteram* 7.18.24.
222. Tertullian, *de Anima* 15.

the functions and interrelations of the soul to include sensation, motion, and memory of motion in a manner similar to and expanding on Tertullian.[223] Augustine states, "It is not the body that perceives, but the soul by means of the body."[224] This implies the self must be stood to one side if God's Spirit is acting through the person.

Augustine uses the same parallels between visions and how the Holy Spirit's agency within other *ekstasis* states. These include comparisons with the state of the disembodied spirit in death,[225] in sleep[226] with inspirational *ekstasis*.[227] Augustine's treatment of Tertullian favours careful analysis and criticism over polemic. "He (Tertullian) was intelligent, he sometimes saw the truth" is typical of this change in tone.[228] Tertullian's scriptural case studies in *de Anima* are re-examined by Augustine. Indeed, Augustine even makes special note of Adam's *ekstasis* when God creates Eve from his rib. Where stupor or sleep is the usual translation, *ekstasis* is the word used in the Septuagint in Gen 2.21.[229] While this is a textually oddity today, it becomes an interchangeable term with sleep or dream language in Augustine's usage.

Augustine did not always acknowledge his positive citations of Tertullian explicitly. For example, according to Taylor "certain writers" is a veiled reference to Tertullian's *de Anima* 5.[230] Taylor and Waszink argue that Augustine's exposition of Adam's sleep in Genesis 2 parallels Tertullian but without reference.[231] What is unexpected is that the structure of the last six books of *de Genesi* show striking parallels with *de Anima*. Augustine's *de Genesi* is much longer than Tertullian's *de Anima*. Therefore as expected, Augustine elaborates more than Tertullian. In book seven of *de Genesi* Augustine follows Tertullian's order in the following topics: that the soul is not part of God's substance;[232] that there is no transmigration of the soul;[233] and

---

223. Augustine, *de Genesi ad Litteram* 7.18.24.

224. Ibid., 12.24.51.

225. Ibid., 12.32.60, 68.

226. Augustine, *de Anima et Eius Origine* 4.27, 28; Augustine, *de Genesi ad Litteram* 12.2.3-4.

227. Augustine, *de Anima et Eius Origine* 4.12.

228. Augustine, *de Genesi ad Litteram* 10.25.41–26.45.

229. *Mentis alienationem* in the Latin version of Genesis used by Augustine. Ibid., 9.19.36.

230. Ibid., 7.21.30.

231. Ibid., 9.19.34; Taylor in Augustine, *de Genesi ad Litteram* 275n95; Tertullian, *de Anima* 11.

232. Augustine, *de Genesi ad Litteram* 7.3.4; Tertullian, *de Anima* 11.

233. Augustine, *de Genesi ad Litteram* 7.9.12–7.10.15; Tertullian, *de Anima* 32–33.

the exposition of medical understanding of the soul.[234] Augustine's criticism of Tertullian's exposition of the rich man and Lazarus[235] is also present but out of order.

Augustine makes two strange departures for a commentary on Genesis. The first follows the discussion of the *ekstasis* sleep of Adam.[236] Augustine strangely moves from a commentary on the text to an extended excursus on the origin and nature of the soul.[237] There is no compelling reason at this point in the text of Genesis for Augustine to move tangentially. Having dealt with Adam's sleep, Augustine states, "But it seems advisable to bring this book to an end at this point."[238] At a similar point in his argument, Tertullian between his references to Adam's *ekstasis* in *de Anima* 11 and 21 also offers a discussion of the origin of the soul, which deals with similar issues in the same order as discussed by Augustine's excursus—traducianism of the soul as the mechanism for the transmission of original sin and the justification of infant baptism.[239] Augustine follows Tertullian at this point rather than Genesis. Augustine's second departure from the text of Genesis is at the beginning of the twelfth book when he moves from a discussion of Adam and original sin to the beginnings of his discussion of the nature of inspiration, visions, dreams and death. Tertullian also has the same transition as he discusses inspiration, visions, dreams and death.[240]

These similarities remain even when the structure of these parts receives closer examination. The similarity of structure between both works is remarkable, leading to the conclusion that Augustine specifically revised Tertullian. While Augustine may treat Tertullian more mildly in *de Genesi ad Litteram* than in other places, there can be no doubt that despite a degree of respect, Augustine heartily disagrees with his predecessor. The lasting significance of their philosophical differences about inspiration relates to the implications they draw regarding the nature of spirit and error in relation to inspiration. These differences shape Augustine's revision of Tertullian.

---

234. Augustine, *de Genesi ad Litteram* 7.13.20–7.20.6; Tertullian, *de Anima* 37–38.

235. Augustine, *de Genesi ad Litteram* 8.5.9; Tertullian, *de Anima* 7.1, 9; Luke 16.24.

236. Gen 2.

237. Augustine, *de Genesi ad Litteram* 10.

238. Ibid., 9.19.36; This is the end of the ninth book. Taylor, following Waszink, argues that Augustine's treatment of Adam's sleep as *ekstasis* parallels *de Anima* 11.

239. Tertullian, *de Anima* 22–43.

240. Augustine, *de Genesi ad Litteram* 12; Tertullian, *de Anima* 44–51.

Inspiration's Highest Expression as Perfect Divine Action

Augustine writes of three levels of reality which reflect decreasing scales of impassibility: the divine or ideal; the spiritual and; the corporeal. Augustine uses these as typical examples: God is in every way unchangeable; the soul is changeable in time but not in place; and the body is changeable in both time and place. This is a frequent theme in Augustine's work and is fundamental in his metaphysics.[241] Augustine's classification of visions parallels the content but revises Tertullian's discussion of error in *de Anima*, in which Augustine applies his three levels of reality to Christian experience of the Holy Spirit as well as his notions of the incorporeality and impassibility of the soul. There are thus, three kinds of vision: intellectual or divine, spiritual and corporeal, where the former is the highest.[242] Each corresponds to a level of reality and a more extensive state of *ekstasis* within the person's mind; corporeal and spiritual visions can err but not the highest, intellectual visions.[243] Augustine's view differs from Tertullian for whom visions were error-proof, but their recall, interpretation and application might introduce error.

While the kinds of *ekstasis* experienced in the life of the Church community are a type of vision similar to corporeal,[244] Augustine argues that Scripture always involves the higher intellectual vision. Augustine explicitly makes a point about using the Latinized form of the Greek *ekstasis* making this point. In this case, Augustine understands the whole mind to have been moved aside while God works in the person perfectly so that God's actions are perfectly enacted and recorded. In this way, Augustine formalises the treatment of Scripture as a special case. Scripture is for him trustworthy is because it is inspired, for by inspiration God guarantees God's own veracity. Note that this derives logically from his description of inspiration within the three levels of reality. In this way Augustine formalises the link between perfect divine action and *ekstasis* inspiration, noting that the anthropology that implies that *ekstasis* is both possible and necessary is derived primarily from the extra-biblical sources of Greek philosophy and medicine.

Anything true related during inspiration is purely God's action. Further, Scripture is especially the sole work of God. "Since they were men who wrote the Scriptures, they did not shine of themselves, but 'He was the true light, who lighteth every man that cometh into the world.'"[245] Augus-

---

241. Augustine, *de Genesi ad Litteram* 8.20.39; Taylor in Augustine, *de Genesi ad Litteram* 2:261n97.

242. Augustine, *de Genesi ad Litteram* 12.7.16, 11.22, 12.9.25–26.

243. Ibid., 12.25.52—12.26.53-54.

244. Ibid., 8.25.47.

245. Augustine, *Gospel of John* 1.6; *Psalms* 116.8.

tine's view of inspiration is so high that under inspiration the Septuagint translators had licence to alter the original Hebrew.

> In the Hebrew there is said to be a different expression: giants being used where physicians are here: but the Septuagint translators, whose authority is such that they may deservedly be said to have been interpreted by the inspiration of the Spirit of God owing to their wonderful agreement, conclude, not by mistake, but by taking occasion from the resemblance in sound between the Hebrew words expressing these two senses, that the use of the word is an indication of the sense in which the word "giants" is meant to be taken.[246]

This helps explain why he was so persistent in seeking a reply from Jerome on his implication that there might be a deliberate simulation in the text of Galatians.

Augustine and Jerome

Augustine was deeply concerned that Jerome could suggest that Paul simulated his rebuke of Peter in relating the account of his dispute with Peter recorded in Galatians. This led to a heated if somewhat protracted exchange of letters between the two. If Jerome's suggestion was the case, then Paul committed a falsehood, but in Augustine's description of inspiration divine agency would in the case of truly inspired writing not contain falsehood.

> For it seems to me that most disastrous consequences must follow upon our believing that anything false is found in the sacred books: that is to say, that the men by whom the Scripture has been given to us, and committed to writing, did put down in these books anything false.[247]

If there is falsehood at this point then the whole of Scripture might be false.[248] Woe betides anyone who suggested that Scripture may contain imperfections. This was Jerome's fate. Augustine indicated that Jerome had better revise his thinking.

> For if you apply more thorough attention to the passage, perhaps you will see it much more readily than I have done. To this more

---

246. Augustine, *Psalms* 88.9.
247. Augustine, *Letters* 28.3.
248. Kelly, *Jerome*, 217–20, 263–72.

careful study that piety will move you, by which you discern that the authority of the divine Scriptures becomes unsettled.[249]

Augustine's other concern was that Jerome had stopped marking changes in wording between the Septuagint and the Hebrew in translating the Old Testament. He wanted an explanation as he saw the Septuagint as the inspired, hence authoritatively perfect translation.[250] With his high view of the Septuagint, Augustine absolutely needed to know how the new Latin translations differed from what he saw as the inspired Greek. Jerome only saw this as hard work and did not see it as important. In a later letter Augustine highlighted what he saw as the pastoral peril in applying Jerome's translations.

> A certain bishop, one of our brethren, having introduced in the Church over which he presides the reading of your version, came upon a word in the book of the prophet Jonah, of which you have given a very different rendering . . . Thereupon arose such a tumult in the congregation, . . . The man was compelled to correct your version in that passage as if it had been falsely translated, as he desired not to be left without a congregation.[251]

Augustine persisted, writing no less than three times to get a reply from Jerome.[252] Kelly details the controversy surrounding the letters, including the delayed deliveries, questioned authenticity, public leaking of Augustine's letters and the confused and acrimonious exchanges with Jerome.[253] When Jerome's reply finally came, it was blunt.[254] Jerome follows Origen who while maintaining a high view of the Septuagint acknowledged the need for interpolation to arrive at the best sense of the text.

> Do you wish to be a true admirer and partisan of the Seventy translators? Then do not read what you find under the asterisks; rather erase them from the volumes, that you may approve yourself indeed a follower of the ancients. If, however, you do this, you will be compelled to find fault with all the libraries of the Churches; for you will scarcely find more than one Ms. here and there which has not these interpolations.[255]

---

249. Augustine, *Letters* 28.3.5.
250. Ibid., 28.2.
251. Ibid., 70.3.5.
252. Ibid., 28, 40, 71.
253. Kelly, *Jerome*, 217-20, 263-72; Augustine, *Letters* 75.
254. Augustine, *Letters* 75.
255. Ibid., 75.5.19.

Jerome suggests that he might have quoted earlier and more extensive commentaries and could not see anything wrong with his explanation as it fitted what he knew as the tradition of interpretation of Galatians. To this letter Augustine finally replies he too is working in his own tradition of interpretation of the passage.[256] While the tone of this letter seems conciliatory, Augustine confronts Jerome with the following:

> For I confess to your charity that I have learned to yield this respect and honour only to the canonical books of Scripture: of these alone do I most firmly believe that the authors were completely free from error. And if in these writings I am perplexed by anything which appears to me opposed to truth, I do not hesitate to suppose that either the Ms. is faulty, or the translator has not caught the meaning of what was said.[257]

On the question of differences between the Hebrew and Greek texts, Jerome suggests a corruption of the original Hebrew in the Septuagint translation.[258] Augustine acknowledges that Jerome's would be a better proposition, but then asks for a copy of Jerome's Septuagint and an explanation of the alterations.[259] Given the acrimony and confusion of the correspondence, Augustine probably was not offering an olive branch to Jerome as it is unlikely that he had actually revised his opinion of the Septuagint.

Kelly interprets Jerome's reluctance to answer the later letters in this correspondence as Jerome refusing to admit he incorrectly interpreted Galatians. This does not, however, account for Jerome's comment on the Hebrew and Greek texts. It is more likely that Jerome simply did not see the point of the fuss. At no point does Jerome seem to understand what is theologically at stake for Augustine regarding perfect scriptural inspiration. This leads to the conclusion that Jerome did not connect scriptural textual infallibility to inspiration and divine agency in humans in the way Augustine had. Jerome's response suggests that Augustine fusion of the three notions— inspiration, perfect divine action and a high view of Scripture—was not theologically essential to the explication of divine agency. While it is solely Augustine's explanation which is reappropriated in early modernity, his exchange with Jerome indicates that it is not necessarily the only possible interpretation. This is import as it is the reappropriation of Augustine's description and the notion of divine agency it contains which has implications for developments between natural philosophy and theology into the nineteenth century.

256. Ibid., 82.3.24.
257. Ibid., 82.1.3.
258. Ibid., 75.
259. Ibid., 82.5.34.

2 Summary of Augustine's Description of Inspiration

While drawing on Tertullian, Augustine significantly revised Tertullian's description of inspiration, strengthening the high nature of special inspiration in contrast to other forms. The features of this Augustinian *ekstasis* description include:

1. The separation of soul or spirit from the physical, with the image of God reflected in an impassable element of the human spirit;
2. The assumption of an metaphysical anatomy that assigns functions of reason, judgement and direction to the soul which uses, interprets the senses and memory, and directs the physical body;
3. A foundation which is a synthesis of Aristotelian philosophy, Neo-Platonism and classical medicine, not Scripture;
4. That the inspiring action of the Holy Spirit within the human person creates an *ekstasis* state similar to sleep, extreme fear, and death;
5. The more complete the *ekstasis* the more reliable the inspired action.
6. The most complete state of *ekstasis* is totally reliable.
7. Scripture is reliable and hence must have been written in this special state.
8. Scripture is therefore infallible.

These points encapsulate a way to speak of the manner of divine agency in human beings, which is predicated on human anatomy having a metaphysical soul which the Holy Spirit displaces. When this displacement is total, the implied perfection of this unalloyed divine action dovetails neatly with the seventeenth-century understanding of divine perfections. This description also offered a rationale for affirming the authority of the books of God's revelation. What will be demonstrated in the next chapter is that the understanding of divine agency encapsulated in this description of inspiration became applied to divine agency in the world in general in a manner that complimented and affirmed early modern understanding of divine perfection and nature as a book of God's revelation.

The description implies that the production of Scripture is a special and best case of inspiration. There is however, no evidence or tradition in the early Church recording the writing of Scripture occurring in special states of inspiration. That is with the exception of the legend of the translation of the Septuagint. The assumption that the writing of Scripture involved special *ekstasis* only arises after Augustine's description. Sasse described

Augustinian inspiration as "A heathen theory taken over via the Synagogue, and which was only superficially given a Christian appearance."[260] He also calls it a disaster.[261] Sasse's strongly worded comments acknowledge the problems that have occurred in the west. On the evidence presented, the *ekstasis* description of inspiration is not specifically Christian in its derivation, beginning instead an understanding of human anatomy based in Greek philosophy and medicine. This allowed it to be readily used in the seventeenth-century environment of generic theology without reference to Christology. The seventeenth-century usage of inspiration saw a reversal in the relationship between the authority of Scripture and inspiration. Prior to this the Scriptures' authority depended on the testimony of the apostolic Church implying their inspiration. Inspiration could become a guarantee of accuracy only when Augustine's description had been taken for granted along with the seventeenth-century understanding of the divine perfections along with the notion of the two books as an answer to the question of authority among Protestants.

Tertullian's description of inspiration makes *ekstasis* mandatory for inspiration, thus making the inspired action error-free except in how it is recalled. However, Augustine asserted pure or total *ekstasis* as the highest form of inspiration and is therefore perfect free of human error even in recollection. While Augustine's view has dominated western theology, surely an alternative description is possible—one not having the same implications whist remaining consistent with Trinitarian theology. Such a description arguably may constitute a more fruitful basis for theological development and ultimately allow a more productive interaction between theology and science.

Incarnational Divine Agency and Inspiration

The earlier discussion of Augustine and Jerome suggests that there already existed, although undeveloped, an understanding of inspiration which does not presume perfect divine action in the same way as Augustine's description, without necessarily rejecting a very high view of Scripture. What can be developed is a description that further does this as well as not presuming the existence of the soul, a particular set of functions of soul or *ekstasis* as the agency of divine action within the soul. The elements of such a non-Augustinian theological anthropology can be identified within the theological writings of Eastern scholars including Athanasius, Leontius of Byzantium

260. Sasse, "Inspiration and Inerrancy," 70.
261. Ibid.

INSPIRATION, PERFECTION, AND GENERIC THEOLOGY

and John of Damascus. Although it is possible to elicit an alternative description of inspiration, it is not as well developed as Augustine's description. It will require development in order to warrant serious consideration as a replacement to the Augustinian description. Such reconsidered divine agency would not lead to the unresolved issue, the impasse that is the focus of this book.

Some late twentieth-century western theologians have argued that the inspiration in writing of Scripture has parallels with the divine-human union in Christ. One such example made by Alonso-Schökel.

> Finally, we should remember that just as the Incarnation is a transcendent mystery to be adored in grateful silence, so, too, inspiration pertains to the realm of this same mystery. Thus, when we inquire into the fundamental problem in the mystery of inspiration and ask, "How can words be at once both divine and human," the answer is spontaneous: in a way similar to that by which Christ is both man and God.[262]

Alonso-Schökel argues that all the traditional views of the Bible have committed the same error. Post-Reformation views reduce the use of language in Scripture to merely the imparting of technical propositions. The assumption is that God has spoken, thus God has produced a series of propositions.[263] This is not so, he argues, for "God has rather assumed all dimensions of human language."[264] Inspiration must of necessity include divine-human activity writing of the original text, its editing, transmission, hearing, preservation and the determination that it is canonical. While Alonso-Schökel at this point refers initially to the inspiration of the words of the scriptural text as a special case, ultimately this should not be divorced from the question of inspiration in general.

What is to be proposed is an incarnational description of inspiration, in which this noted similarity actually derives from the incarnation. It will be shown to relate to God's renewal of humanity in Christ by the agency of the Holy Spirit. This description will be developed beginning with the Holy Spirit's recreating work of communicating that renewal to humanity. Gunton argues for such a recreation of the creature by God, deriving this from the incarnation.[265] The dynamic of God's recreation of the creature

---

262. Alonso-Schökel, *The Inspired Word*, 53.
263. Ibid., 56, 96.
264. Ibid., 325.
265. Gunton, *Christ and Creation*, 35–59, 106–27.

is by God's self-communication of renewed humanity in the humanity of Christ. This communication is made in and by the Holy Spirit.[266]

Application of *Anhypostasia* and *Enhypostasia* to Inspiration

Gunton grounds the Holy Spirit's work in the renewal of humanity in the incarnation of Christ,[267] by reference to the individual Eastern doctrines of *anhypostasia*[268] and *enhypostasia*.[269] Christology and Pneumatology are inexorably linked through using of these two notions, according to Gunton. Barth also argues in the strongest terms that *anhypostasia* and *enhypostasia* must be linked "s God cannot be considered without His humanity, His humanity cannot be considered or known or magnified or worshipped without God. Any attempt to treat it *in abstracto*, in a vacuum, is from the very first a perverted and impossible undertaking."[270]

These doctrines were developed to defend the Chalcedon creed's definition of Christ's "two natures" against heresy: *anhypostasia* against that of the Monophysite Severus[271] and *enhypostasia* against that of Nestorius.[272] Together these christological formulae argue for the preservation of the Chalcedon "one person, one hypostasis, two natures"[273] as the orthodox middle ground between the heretical extremes of the Monophysite's "one person, one hypostasis, one nature"[274] and the Nestorian's "one person, two hypostases, two natures."[275] The doctrines of *anhypostasia* and *enhypostasia* argue for the preserving union of both natures, divine and human. The term *anhypostasia* refers to Christ's humanity having no substance or "personality"[276] of its own apart from the divine nature—that is, Christ is

---

266. Ibid. Also, see Torrance, *Theology in Reconstruction*, 192–99.

267. Gunton, *Christ and Creation*, 46–52. In this he appears to draw on Torrance, "Arnoldshian Theses."

268. Leontius, *Adv. Severum* (trans. Daley, "The Origenism of Leontius of Byzantium," 338).

269. Leontius, *Liber Tres contra Nestorianos et Eutychianos*, PG 86a:1367–95, 1277–80 (trans. Grillmeier, "The Christology of Leontius of Byzantium," 194).

270. CD IV/4:102.

271. Florovsky, *The Byzantine Fathers of the Fifth Century*, 168.

272. Rees, "Leontius of Byzantium and His Defence of the Council of Chalcedon," 111–9.

273. Chapman, "Monophysites and Monophysitism,"

274. Ibid.

275. Ibid.

276. Schaff, *The Creeds of Christendom*, 32; Alternatively "Christ's human nature . . . lacking a hypostasis of its own." Lang, "Anhypostasis-Enhypostasis," 656.

truly human, without putting on or subsuming an independent human nature. Christ's nature is human nature and not a copy or a replica of it following the principle "that which he has not assumed he has not healed."[277] *Enhypostasia* notes that this humanity of Christ and human nature is further eternally preserved within God.[278]

The notions of *anhypostasia* and *enhypostasia* were formulated to address issues in Christology. They were not originally applied to Pneumatology or inspiration by the post-Chalcedon Eastern Fathers who used them. Indeed, there is considerable debate about who actually developed the notions in the seventh century and that they may not have been paired as concepts until the sixteenth.[279] The present discussion offers an opportunity for such application.

The conventional *anhypostasia* will be used, even as highlighted by this debate, it is a mistransliteration of *ahypostasia*. Modern discussions tend to use noun and adjectival forms of both words indiscriminately.

Anhypostasia was coined to counter the Monophysite assertion that Christ could only have one nature, the divine, thus subsuming the human nature; the divine nature subsumes at least part of Christ's humanity making it less than fully human. In the Monophysite understanding, it is impossible for both natures to coexist. The *ekstasis* description of inspiration in effect argues for a similar inability of the Holy Spirit and human spirit to coexist. Jesus' human spirit with its consciousness must thus be stood to one side as the Holy Spirit acts in inspiration.

It is possible to apply *anhypostasia* and *enhypostasia* to inspiration as follows. The Holy Spirit given by the Father and received by Christ into Christ's humanity preserving the union of divine and human natures. The preservation of the two natures does not simply result from Christ's divine action. By *anhypostasia*, this action cannot be separated from being at the same time a fully human action. Christ's action in receiving and depending on the Holy Spirit must be both fully human and fully divine. However, no human action can contain God. It is only in inaction by *kenosis* or willingly self-emptying, which Christ chooses in non-action to be utterly dependent

---

277. Gregory of Nazianzus, "Epistle 101," 218.

278. Gunton, *Christ and Creation*, 46–52; Watson, "A Study in St Anselm's Soteriology and Karl Barth's Theological Method," 493–512.

279. Shults, "A Dubious Christological Formula: From Leontius of Byzantium to Karl Barth," 431–36; Lang, "Anhypostasis-Enhypostasis: Church Fathers, Protestant Orthodoxy and Karl Barth," 630–58; Daley, "A Richer Union: Leontius of Byzantium and the Relationship of Human and Divine in Christ," 239–65; Daley, "The Origenism of Leontius of Byzantium"; Grillmeier, "The Christology of Leontius of Byzantium." Lang, "Anhypostasis-Enhypostasis," 656–57.

on the Holy Spirit received into his humanity, to be the agency which preserves the union and enables the renewal and recreation of humanity in relationship with God. The Holy Spirit's action in other humans derives from this unique action of the Holy Spirit in the humanity of Christ. It is by the Holy Spirit received into his flesh and shaped by his *enhypostatically-preserved* humanity that Christ continues to act in people to unite them to God and renew them as people.

The union of the Holy Spirit with Christ's humanity constitutes the manner of this divine agency in human beings. This union mediates and shapes the interaction between the Holy Spirit and their sinful humanity. This action of the Holy Spirit preserves and renews sinful humans whilst being shaped by the interaction between the sinless humanity of Christ and the Holy Spirit in Christ. This goes towards answering Torrance's question, "Why did Christ receive the Holy Spirit in his obedient life and self-sanctification?"[280] Christ, being fully God, did not need the Holy Spirit to do this. He did this on our behalf, to recreate our humanity in his human body by the Spirit. It is this Spirit in Christ's humanity who is poured out on the Church at Pentecost to enable all Christians to participate in the recreation of their humanity in Christ.

Further, *enhypostasia* points to the eternal preservation of the humanity of Christ within the eternal being of God, outside or independent of created space-time. The agency of the Holy Spirit within Christ's humanity while occurring in history is by God's choice not bound to it, transcending time in eternity. This suggests how personal divine agency might work historically before Christ.

Generally, humans do not receive the Holy Spirit in the same fashion as Christ; rather Christ becomes the special case which grounds and forms humanity's experience of God's renewing work. Inspiration is thus grounded in and derived from Christ's divine-human nature. Thus, the manner of divine agency giving inspiration comes to depend upon who is its foundational mystery—Christ. Inspiration is an act of Christ by the Holy Spirit in a person in which God communicates God's self. This depends on and is shaped by the incarnational union of the divine and human in Christ, be it for the production of Scripture or for the preached word or for personal or corporate illumination. The inspiration of Scripture becomes one specific case of this work rather that a different kind of action. It remains special not because it is an action perfected in degree, but because the tradition of the church indicates that it is the standard against which all other acts of inspiration are measured.

---

280. Torrance, *The Mediation of Christ*, 192.

This description of inspiration does not presume that divine action be perfect. Divine agency of the Holy Spirit within the human person allows for both the preservation and a process of reshaping the human nature in the image of Christ. Perfected divine action might be the result of this change but is not an automatic result of the Holy Spirit's presence in the person.

Summary of Incarnational Inspiration

In contrast to the Augustinian description of divine agency, the incarnational description as suggested makes these opposite points:

1. God's choice to act in the world and humans does not depend on any non-theological description of what humans or humanity might be;
2. The agency of the Holy Spirit in humans is shaped and derives from Christ's continuing reception of the Holy Spirit in his *enhypostatic* humanity without assuming a metaphysical anatomy or a particular relationship or distinction between the human soul or spirit and the physical;
3. Inspiration depends on the central mystery of the incarnation rather than on any preconceived philosophical, scientific or medical ideas;
4. The Holy Spirit's action does not automatically require *ekstasis*;
5. The Holy Spirit agency can occur during a broad range of human activities or emotional states in which the Holy Spirit fully preserves the person's humanity as the Holy Spirit acts;
6. The writing of Scripture need not be considered to have occurred by a different class of activity conducted under the inspiration of the Holy Spirit;
7. Inspiration by divine agency does not automatically guarantee that perfected human action is a result.

This description does not start by assuming knowledge of what God is or of humanity is assumed to be. It is essentially different to that developed by Augustine. This will become important in the chapter after the next in which the Augustinian description's assumptions of a metaphysical soul and perfect divine action will be shown to become problematic for debate between theology and science in the mid-nineteenth century.

What has been demonstrated in this section is that initially the question of divine agency and inspiration in particular are broader issues within Pneumatology rather than merely being an aspect of the doctrine of Scripture. Tertullian in the process of explaining the Holy Spirit's agency within his culture incorporated revised Aristotelian ideas and medical knowledge in his description of inspiration. Augustine appears to have directly revised Tertullian's description further cementing the notion of a metaphysical human anatomy within a description which requires such divine agency or inspiration to automatically require a standing aside of this metaphysical soul. In its ideal expression, this *ekstasis* is total and the resulting inspired activity is without error or perfectly God's own action.

The task of the next two chapters is to trace how the *ekstasis* description of inspiration gave rise to an understanding of divine agency which:

1. helps to shape understanding of the natural world;
2. drew a parallel between divine agency in humans and in the world;
3. helps to shape the understanding that the natural world obeys intelligible laws;
4. helps to shape the understanding that those laws are perfectly inspired by God in nature; and
5. led to a detailed study of the world called that perfection into question.

Demonstrating these influences in the development of modern science will lead to showing how an understanding of divine agency arising from the three factors discussed eventually became one of a possible number of stumbling blocks to the dialogue between theology and science. Once this progression has been traced the next step will be to examine whether the proposed incarnational description offers an adequate alternative way forward in its describing divine agency christologically and pneumatologically. Does altering this description and thereby the shape of inspiration, as one of the three factors discussed, help to avoid one impasse that has developed in the dialogue between theology and science? Examining the incarnational description in conversation with the theology of Barth will determine whether this description of divine agency is coherent or robust enough to become part of a revised basis for the dialogue between theology and science.

It is useful to locate both Newton and Darwin in relation to particular aspects of historical developments involving the legacy of Augustine's *ekstasis* description of inspiration, the notion of the two books and early modern understanding of divine perfection. This will justify the choice of using case studies over an extended survey of this period of history. The changes in thought between Newton, Darwin and Huxley are indicative of key changes and trends during this period as well as being profoundly influential in their own right. The set of circumstances which will be argued leads to the impasse relies on there being little change in the shape of the three factors giving rise to how divine agency was understood. It will be sufficient to demonstrate their continued status and their key roles in each of these cases.

Appleby, Hunt and Jacob extensively illustrate the importance of the work of Newton, Darwin and Huxley in the development of modern and post-modern thought.[281] Lash, while outlining what he describes as the development of dualistic patterns in modern thought also described the influence of the debate between Newton and others on eighteenth-century transitions in theology and natural philosophy.[282] Frei pointed to the rise of Newtonianism in the eighteenth and Darwinism in the nineteenth Centuries, as significantly affecting the nature of interpretation.[283] Chadwick devotes a whole chapter to the place of Darwinism in the secularisation of Europe in the nineteenth century.[284] Barth devotes large sections of chapters 41 and 42 of *Church Dogmatics* to implications raised by Newton's protagonists Leibniz and Déscartes in relation to the doctrine of creation. Developments during the lives of Newton and Darwin in relation to the development of modern thought have attracted much recent scholarly attention. There is little doubt that the reappropriation of Augustinian theology had a significant influence on the development of thought during these centuries. Nevertheless, the place of Augustine's inspiration, its assumptions and its corollaries on the development of the understanding of divine agency remains an open question.

---

281. Appleby, Hunt, and Jacob, *Telling the Truth About History*, 198.
282. Lash, *The Beginning and the End of Religion*, 2–25.
283. Frei, *The Eclipse of Biblical Narrative*, 51–65.
284. Chadwick, *The Secularization of the European Mind in the Nineteenth Century*, 189–228.

CHAPTER 3

Newton and God/Providence Inspiring the Universe

THE CASE HAS BEEN argued for the significance of three factors in early modernity's development of an understanding of divine agency in the world. The application of the two books of God's revelation had led to a confidence that nature could be understood rationally and that disciplined investigation could reveal more about God and God's purposes. This in turn was supported by the widely held understanding that God's perfections would be reflected in the world through nature being characterized by goodness, simplicity and elegance and that everything that exists must have a creator-given purpose. Ultimately nature was seen as being harmonious and being so well ordered so as to not require revision. Evil where it exists was understood to be on the balance only that which permitted the best good to be developed in general. The third factor, inspiration, it has been argued, became a guarantee for the authority of divine revelation for both the veracity of divine law written in Scripture. It also was understood as God's necessary assistance for natural philosophers' disciplined and rational investigation of the natural world. This Augustinian *ekstasis* description of inspiration contains an understanding of divine agency in humans which overcomes ordinary human limitations and error. The reliability and veracity of this revelation by such inspiration was not disputed even when its interpretation by the church or even its accurate transmission in the Scriptures was doubted. The way these three factors were expressed in the seventeenth century allowed the description of God to be made by reference to general principles and without specific reference to whom God may have specifically revealed God's self to be in Christ. While this non-christological approach was itself a matter of dispute, it demonstrably allowed people with a variety of understandings about who God is to become allies defending religious belief. This however, becomes more important in the nineteenth rather than the seventeenth century.

While the notion of the two books led to the expectation that what God had revealed in nature could also be as reliable as that revealed in Scripture, what had and still has not been established is the agency by which God writes or maintains these laws in nature. What this chapter will establish is that a parallel was derived between divine agency in humans as described in *ekstasis* inspiration and divine agency in the world. This is evident in the work and influence of Isaac Newton.

There are notable problems in discussing Newton which should not be underestimated. Technical issues of historical bias can be identified in a long tradition of revisionist and politicized histories in which a narrow focus on Newton's more "respectable" studies has led to the neglect of the breadth of his thought and even to the long loss to scholarship of valuable primary documents. By taking account of these biases and in using newly available primary sources it is possible to show that Augustine's *ekstasis* description of inspiration had a foundational role in Newton's understanding of the natural world and his theology. Further, influence of the other two factors is present both in his own work and in the assumptions that he and his protagonist Leibniz leave unchallenged in their thinly veiled but heated exchanges under the mediation of Clarke. Ironically even though they rarely agreed on anything, both made similar assumptions about human metaphysical anatomy and the actions of a divine perfect-being which relate the understanding of divine agency in the world. In considering their theological assumptions, there is no doubt that their thought had implications for the development of philosophy, theology and science in the centuries following.

Newton as Theologian

Surveys of Newton's theology usually refer to the *Scholium* added to the later editions of Newton's *Principia* and to the debate on the nature of God conducted by Clarke with Leibniz. Newton's theological presuppositions are identifiable in this debate which Clarke conducted on his behalf. However, Newton wrote far more on theology than he did in relation to science, only a small selection of which he ever published.

The great bulk of Newton's writings on religion, an immense volume of papers running to several million words ... Some of them are notes from his reading. He started with the Bible; twenty-five years later, John Locke would confess that he had never met anyone with a deeper knowledge of the Scriptures. From the Bible, Newton proceeded on to the early Fathers of the Christian church. Again he was nothing if not thorough; he read extensively in the works of such men as Origen, Athanasius, Gregory Nazianzen, Justin Martyr, and Augustine. If we can be guided by the notes he left, Newton became as exhaustively familiar with patristic literature as he was with Scripture.[1]

A difficulty for Newton-studies until recently has been that most researchers have not had ready access to a large part of his writings excepting his better known scientific works and the volumes of his correspondence. Fara has traced the "almost farcical air" of the disposition of the intestate Newton's papers after his death, including the deliberate exclusion of bundles of papers from published works. Exclusion typically was based on what was deemed not appropriate for Newton's eulogising in a given period.[2] For example, his extensive alchemical works are largely inaccessible and still await detailed scholarly analysis. Newton's *de Gravitatione* was published for the first time in 2004, nearly three centuries after Newton's death.[3] Popkin and Iliffe have described in detail the challenges social, academic and political involved in making these available.[4]

As a result, it has been difficult to gauge the breadth and full influence of his ideas. Fara has investigated how the long history of the mythologising distortion of the man and his work further complicates the task of understanding his influence within history[5] and even in contemporary research.[6] Many of his unpublished opinions were known to his associates and influenced the ideas of those who followed him. Consequently, the heterodoxy of his theology was well if not widely known. What is surprising is that in a period not noted for religious toleration, Newton received royal preferment, being exempted from ordination in order to hold the Lucasian chair,

1. Cohen and Westfall, *Newton*, 328.
2. Fara, *Newton: The Making of Genius*, 27–29.
3. Janiak, *Isaac Newton: Philosophical Writings*, 12–39.
4. Iliffe, "Digitizing Isaac: The Newton Project and Electronic Edition of Newton's Papers," 23–38; Popkin, "Plans for Publishing Newton's Religious and Alchemical Manuscripts, 1982–1998," 15–22.
5. Fara, *Newton: The Making of Genius*.
6. Jacob, "Introduction," ix–xvii; Osler, "The New Newtonian Scholarship and the Fate of the Scientific Revolution," 1–14.

in becoming master of the mint, and also finding prominent placement for protégés such as Clarke and Whiston.

A complex Original Thinker Rather than a "Simple" Scientific Hero

There is a common and unhelpful view of Newton as the heroic pioneer of science exemplified in the verse from Alexander Pope:

> Nature, and Nature's Laws lay hid in Night.
>
> God said, Let Newton be! And all was Light.[7]

Even after the heroic myth of Newton, the stoic, virtuous, calmly rational independent man of science is rightly subjected to healthy iconoclasm in contemporary scholarship, a number of intriguing questions remain about his theology and his science. These questions relate to the purpose of the present analysis, which will demonstrate a link in Newton's drawing a parallel between divine agency in humans contained in inspiration and divine agency in the world.

Recent trends in Newton scholarship highlight the complexity of the man who, while without question brilliant in his natural philosophy and his explanations of God's laws in nature, did not fit a stereotypical myth of scientific genius. Because Newton was a complex person it is useful to resist simplistic labelling of his thought. Jacobs, Appleby and Hunt have argued particularly in the case of Newton, the myth of the heroic scientist, while having been a dominant explanatory paradigm, has not encouraged either an engagement with the complexities of the early scientists or an appreciation of them as people engaged in the debates and philosophical questions of their day.[8] Under the "heroic scientist" paradigm, it was usual to note Newton's interest in theology as peculiar, and his interest in alchemy even more so. Newton's loathing of criticism and notoriety for not submitting his research and ideas to public scrutiny also does not fit the paradigm. Nonetheless, he remained deeply influential in the academic life of his day and in the centuries following. He was a man of insightful reason but not given to diplomacy with those whom he considered rude, impolite or simply wrong. As Iliffe has noted, Newton fell short of faultlessness with "his often graceless treatment of both friend and foe."[9]

---

7. Fauvel et al., *Let Newton Be!*
8. Appleby, Hunt, and Jacob, *Telling the Truth About History*, 29, 42–43.
9. Iliffe, *Newton: A Very Short Introduction*, 5.

The Scripture Scholar searching for pure unadulterated revelation

Arguably a useful and comprehensive interpretative key to Newton's theology and its influence on his natural philosophy is to begin with his understanding of Scripture. The apparent strangeness of Newton's theology, his commitment to finding alchemical truth and to describing order in nature, it will be shown, all derive from his acceptance of the idea that divine action was perfect, necessarily implying flawless revelation. This, in turn, will be shown to depend on the notion of divine agency in Augustinian *ekstasis* inspiration. This complements the notion of the two books and the implications of perfect-being theology for the created world.

Newton's notes show evidence of his demonstrated familiarity and agreement with specific aspects of the Augustinian description of inspiration both in general and in detail. Newton's commitment to the perfection of divine action resulting in a perfectly revealed text of Scripture is demonstrable in the value Newton placed on Lucius Cary's[10] work. Newton's extensively used commonplace book, the Theological Notebook, shows a list of 73 historical authors, some annotated with "Trin. Coll." and shelf number. There is a list of "*Authores Notandi*" indicating what Newton believed to be influential works. The first in this list is the set of papers entitled Lord Falkland's *Discourse on Infallibility, Out of the Jesuit's Answer, The Lord Falkland's Reply* and *Out of Lord Falkland's Reply*.[11] Newton summarises Cary's arguments over six pages, defending scriptural infallibility against "papist" arguments ascribing authority to the Church rather than Scripture.[12] Newton saw the link between Augustine, Tertullian and scriptural infallibility as the basis for Scripture's authority in determining matters of faith: "All other opinions witnessed by any other ancients to have tradition may have been by them mistaken to have been so, out of St Austin's & Tertullian's rules: whereas for this, & for this alone are delivered the very words which Christ used when he taught it."[13] Newton's notes demonstrate his detailed familiarity with many of the Church Fathers. He refers to both Tertullian and Augustine frequently throughout the notebook. Newton shows significant sophistication regarding the texts of the Fathers. In particular he criticizes attempts to edit Tertullian. This supports the case made for the consideration of inspiration

---

10. Lord Falkland.

11. Cary, *Discourse of Infallibility*; McLachlan, *Sir Isaac Newton: Theological Manuscripts*, 127. Three papers—*Discourse on Infallibility, Out of the Jesuit's Answer, The Lord Falkland's Reply* and *Out of Lord Falkland's Reply*—were later published in one volume by the Lord Falkland as, *Discourse on Infallibility* (1651).

12. Newton, *Theological Notebook*, Cambridge Keynes Ms.2., 2–7.

13. Ibid., 2, 10.

in the previous chapter. Newton like his contemporaries did not conceive of any other description of inspiration other than that described by Augustine.

A common dismissal of Newton simply describes him as Arian. This is a simplification which misses significant detail in his thought, particularly in relation to the nature of divine agency. Newton in his note books, while sympathetic to Arius overtly avoids siding with Arianism in these private notes. Rather, he gives careful thought to highlighting weaknesses in the arguments used against Arius.[14]

Newton places more emphasis on infallibility than Arius. This will be argued to derive from his understanding of divine agency. He paralleled Cary's sentiments, "And though the apostles write not their native Tongues, yet they write in an inspired language, so that they were not likely to commit, at least, any such solecisms as should destroy the end of the inspirer."[15] Newton in his own commentary also wrote,

> John did not write in one language, Daniel in another, Isaiah in a third & the rest in others peculiar to themselves; but they all wrote in one and the same mystical language as well known without doubt to the Prophets as the Hieroglyphick language of the Egyptians to their Priests.[16]

Newton's commentary on Daniel and the Apocalypse assumes by virtue of inspiration that there is no error in the use of language or metaphor in the text. "For God has so ordered the prophecies, that in the latter days the wise may understand, and none of the wicked will understand."[17] "For understanding the prophecies, we are, in the first place, to acquaint ourselves with the figurative language of the prophets."[18] Newton was a literalist and logically methodological in his approach to Scripture, believing that interpretation of the writings in Scripture considered prophetic must be in relation to their figurative meaning. While this prophetic language may be figurative, it had, according to Newton, one true meaning. Newton used a fixed signification of a prophetic metaphor along the lines that "as agrees best with all the places" of its use, as a principle of interpretation.[19] The meaning of Scripture and particularly of prophetic writing would become clear if one found the correct signification of prophetic metaphor.

14. Ibid., 19–21.
15. Cary, *Discourse of Infallibility*, 184.
16. Newton, Keynes MS 5:1, in Hutton, "More, Newton and the Language of Biblical Prophecy," in Force and Popkin, eds., *The Books of Nature and Scripture*, 48.
17. Newton, *The Prophecies of Daniel and the Apocalypse*, 1.
18. Ibid., 2
19. Newton, "The Language of the Prophets," 120.

There is a further step of perfection to this high idea of the perfection of language. Why would God's earliest revelation not have been totally perfect? It follows, if God reveals perfectly, that the revelation of God must always have been perfect and complete. If what went earlier is perfect and complete, then how could the incarnation of Christ possibly add to the perfection of the unadulterated body of revelation already given though perhaps forgotten? Westfall reports that in Newton's unpublished *Theologiae Gentiles Origines Philosophicae*, he "deflated the role of Christ in human history. Christ came to call mankind back to the one true religion and to that religion he added nothing."[20] Therefore, according to Newton, Christ essentially adds nothing to the revelation of God, except maybe to give a key to the better understanding of that previously revealed—this is a very low and unorthodox Christology. Christology was thus incidental to Newton's primary concerns about God's providence and the omnipotence of the creator. Such a view made Jesus different in degree as perhaps the best of created beings. However, this is a crypto-Arian notion as it is based on a different premise to that of the ancient heresy. Newton applied a sophisticated knowledge of patristic scholarship and his grasp of the complexity of post Nicene theology to his rejection of Nicene orthodoxy. He did not simply follow the ancient heresy. Pfizenmaier arguing that Newton's understanding of Patristics must be included in understanding Newton's theology, noted a possible change in Newton's theology in his later life,[21] Newton's Christology remained subordinationist. Snobelen rightly concluded that Pfizenmaier's claim that Newton might even represent a form of post-Nicene orthodoxy does not seem to be sustainable.[22] Newton did not see himself as orthodox. While the label "Arian" may be useful as a simple description of Newton's theological position, it is not useful for investigating the interplay of his ideas. It is a label like Socinian that Newton did not own for himself.[23] The suggestion here is that Newton's heterodoxy does not stem from his formal adherence to one of the established heterodox schools, but rather, as will be presently shown, to his own resolution of what he saw as a paradox. Newton understood the writers of Scripture had been inspired to write perfectly. This is paradoxical in western theology as no human can be free of ignorance or error, except Christ. Newton resolves this paradox on the side of divine inspiration overcoming human error through the choice of God's will.

---

20. Westfall, "Newton and Christianity," in Cohen and Westfall, *Newton*, 367.
21. Pfizenmaier, "Was Isaac Newton an Arian?"
22. Snobelen, "Isaac Newton, Heretic: The Strategies of a Nicodemite."
23. Ibid., 19–21. Newton, "Queries Regarding the Word '*Homoooousios*,'" in McLachlan, *Sir Isaac Newton*; Newton, *Two Notable Corruptions of Scripture* III.

Nevertheless, while holy writ had been once revealed perfectly Newton believed it could be and had been obscured by inaccurate transmission or deliberate distortion.[24] Westfall observes, "Newton's determination to unmask this ancient crime, together with his study of alchemy, absorbed virtually all of his time for fifteen years." Newton claimed in his *Queries* that there had been changes to the legacy of the Church to perpetrate a massive fraud from the time of the fourth and fifth centuries, corrupting them to support trinitarianism. In his opinion this corrupting revision included the Scriptures. Newton, the master of the Royal Mint who sent many a forger to the gallows, used his considerable legal ability to prosecute at length the influence of Athanasius on the propagation of the Nicene creedal formula[25] and also argued strongly against the *homoousios* of the Nicene Creed.[26] His distrust of a supposedly highly immoral Athanasius depended on descriptions in the works of pseudo-Dionysius. While now known to be a late forgery, in Newton's time they were thought to be authentically ancient. Given his very high view of the original textual perfection of Scripture, this helps to explain his search by other means for an unadulterated divine text as the basis for his theology.

Newton had a commitment, therefore, to finding the original perfect revelation either in the earliest text or in other traditions which had intellectual weight in his day. It was thought that it might be possible to find an ancient unadulterated and possibly ultimate divine message in alchemy and in the related presumed tradition of the Hermes Trismegistus and a related tradition referring to the possible existence of an encyclopaedia of Adam's knowledge.[27] Finding such a perfect message in the language of angels or in some primevally revealed perfection in alchemy is consistent with Newton's commitment to discovering pure perfect divine revelation. It is a continuation of, rather than (as Westfall has tentatively suggested) a "rebellion" from, his earlier studies in mathematics and natural philosophy.[28] Newton's understanding of what he was doing in natural philosophy was to provide further keys for unlocking earlier revelation, the same role as he gave to that of the prophets and scholars in scriptural study and even to Christ himself.

24. Newton, "Paradoxical Questions Concerning the Morals and Action of Athanasius and His Followers,"

25. Iliffe, "Prosecuting Athanasius: Protestant Forensics and the Mirrors of Persecution." Cohen and Westfall, *Newton*, 122–23; Manuel, *The Religion of Isaac Newton*, 58; Newton, "Paradoxical Questions Concerning the Morals and Action of Athanasius and His Followers"; Newton, *The Correspondence of Isaac Newton*, 3: letter 358.

26. Newton, "Queries Regarding the Word 'Homooousios,'" 44–53.

27. Harrison, *The Fall of Man and the Foundations of Science*, 17–51.

28. Westfall, *The Life of Isaac Newton*, 118.

Newton Studies: Open-ended and Controversial

Newton research has and continues to raise strong feeling and opinions among academics. While it is clear that a comprehensive understanding of Newton remains incomplete, opinions and theories are held very strongly. Jacob writing in 2004 recalls the publicly hostile response to the presentation of Westfall's early work on Newton's alchemy in the 1970s, that the "very thought of Newton as theologian and alchemist was found to be repellent in some quarters."[29]

Jacob notes that many opinions concerning Newton's work were formed before a definitive compendium of Newton's correspondence was published in the 1960s, at a time when most commentators firmly held a "rationalist definition of science that was alien to what Newton was trying to do."[30] By the 1980s, the importance of Newton's theological works had become realized and research by 2004 had only just begun to yield results.[31] Far from being well explained and contrary to strongly championed views, Newton studies are far from being able to clearly present a comprehensive description of Newton's thought and consequently his later influence. Jacob argues "[r]e-evaluating Newton opens the whole of early modern intellectual history for re-evaluation."[32]

Prevailing myths about Newton among the general academic community cause even the most careful re-evaluation of Newton intense criticism. There is even prejudice from strongly held opinions. Jacob agrees with Popkin's review of Newton studies since 1982 that the influence of political and social factors has held back the publication and study of Newton's alchemical and theological works lest Newton been seen to be anti-rationalist. Popkin argues that "Newton's so-called irrational side makes sense when seen in context and provides much greater understanding of what Newton himself was trying to accomplish in all of his intellectual work."[33] Popkin refers to the recent work of Dobbs, Force, Iliffe, Champion, Mandelbrote, Harrison, Hutton, Osler, Jacob, Snobelen, Principe and Stewart. Principe also notes,

> [C]ontary to the fears of those who once thought it judicious to suppress these parts of Newton's activities, Newton's looming stature in the history of science remains undiminished by

---

29. Jacob, *Newton and Newtonianism* x.
30. Ibid., ix.
31. Ibid.
32. Ibid., x.
33. Popkin, "Plans for Publishing Newton's Religious and Alchemical Manuscripts," 22.

revelation of his "non-scientific" endeavours. While Newton (like any other early modern figure) can no longer be considered the prototype of the hard-nosed rationalist of positivist scientist succeeding generations wanted him to be, we find ourselves with a Newton undoubtedly even more intriguing than before.[34]

Osler, commenting on post 2000 trends in Newton research, has noted the that difficulty in (much of the 1960's Newton research in) interpreting Newton manuscripts came from the imposition of then present understanding of the "terminology of modern science onto Newton's career."[35] Historical and ideological bias is nothing new in histories of Newton. Fara has outlined examples which make study of Newton problematic from the period following Newton's death, through Brewster's bowdlerising of Newton's anti-trinitarianism through to symbolisms empty of any detail of the actual man's life as exemplified in Dali's surrealistic hollow sculpture.[36] Osler specifically refers to presentism in 1960s Newton research in the mathematics and science focused work of Whiteside, Koyré and Burtt.[37] In agreeing with Force, Osler argues that re-evaluation of Newton's work shows that the diverse aspects of Newton's thought show them to be a coherent whole.[38] However, it has to be argued that Osler does not go far enough. While recognising presentism as a historiographical bias in works focusing on Newton's mathematics and science, she fails to identify the same bias in the received wisdom of Foster's voluntarism hypothesis.[39]

Foster's Voluntarism Hypothesis Reconsidered

Foster's voluntarism hypothesis is the label term often used to name a widely used complex argument which Foster developed in three papers in the 1930s on the relationship between theology and the development of

---

34. Principe, "Reflections on Newton's Alchemy in Light of the New Historiography of Alchemy," 205.

35. Osler, "The New Newtonian Scholarship and the Fate of the Scientific Revolution," in Force and Hutton, eds., *Newton and Newtonianism*, 3.

36. Fara, *Newton: The Making of Genius*. Brewster, *Life of Sir Isaac Newton*; Brewster, *Memoirs of the Life, Writings, and Discoveries of Sir Isaac Newton*.

37. Burtt, *The Metaphysical Foundations of Modern Physical Science*; White, *A History of the Warfare of Science with Theology in Christendom*; Whiteside, "The Expanding World of Newtonian Research"; Koyre, "The Significance of the Newtonian Synthesis."

38. Osler, "The New Newtonian Scholarship and the Fate of the Scientific Revolution," in Force and Hutton, eds., *Newton and Newtonianism*, 8.

39. Jacob, "Introduction."

modern science.[40] It is a complex argument that is not always coherent or internally consistent. Nevertheless, this theory is often cited to purportedly explain the influence of Christian theology on the development of modern science. Foster achieved this at a time when this was generally considered beyond the pale.[41] As Brooke and Cantor explain,

> The point here is that a theology that emphasises the freedom of the divine will to make one world rather than another is a theology that makes it inappropriate to reason *a priori* about how the world must be. Empirical methods are necessary to discover which of the many possible worlds the deity might have made has in fact been made.[42]

Foster's hypothesis has been vigorously supported; Harrison warns that it has become "entrenched."[43] Any study of Newton needs to be careful with this hypothesis. Nevertheless, Foster's thesis suffers significant problems. It is necessary to dispute this theory. In his second paper Foster makes a central claim in which he alleges that "the rise of modern natural science depended upon the rejection of the conception of nature as ensouled; . . . this conception of nature is incompatible with the doctrine that nature is created or made."[44] This is simply incorrect as will be demonstrated repeatedly in Newton's own words later in this chapter. Newton's understanding of the way God acts to inspire and sustain the cosmos actually spiritualises the material world. Newton entertained notions the soul or spirit in nature and material objects throughout the length of his career.

The conclusion of Foster's three papers is that a Christian and Trinitarian understanding of the doctrine of creation which believes God made the world as a material artifice has replaced the ancient Aristotelian understanding of a divinely generated essentially spiritual nature. This, he further argues, came about by the admission of elements of voluntarism into a rationalist theology.[45]

---

40. Foster, "The Christian Doctrine of Creation and the Rise of Modern Science"; Foster, "Christian Theology and Modern Science of Nature (I.)"; Foster, "Christian Theology and Modern Science of Nature (II.)."

41. Harrison, "Was Newton a Voluntarist?," 40; Jacob, "Introduction"; Force, "The God of Abraham and Isaac," 185, 194; Henry, "'Pray Do Not Ascribe That Notion to Me': God and Newton's Gravity," 135.

42. Brooke and Cantor, *Reconstructing Nature*, 20.

43. Harrison, "Voluntarism and Early Modern Science," 1.

44. Foster, "Christian Theology and Modern Science of Nature (I.)," 452.

45. Foster, "Christian Theology and Modern Science of Nature (II.)," 5.

Brooke and Cantor liken Foster's hypothesis to the myth of warfare between science and theology: They argue both, "gloss over the diversity and the complexity of positions taken in the past."[46] Harrison also argues that the terms, voluntarism, necessity and contingency have been defined so vaguely as to make the thesis "virtually meaningless."[47] Typically Foster's voluntarism is used to contrast intellectualism and divine necessity with contingency in the natural world.

This begs the question: Why does Foster's hypothesis continue to receive serious attention? Foster's case is weakened by the fact that he makes few references to historical persons or their work. Nonetheless, it seems exceptionally plausible, because Newton's editor Coates writes of God's voluntary action in his introduction to the second edition of the *Principia*. Oakley specifically linked Foster's hypothesis to Newton via Coates' introduction.[48] However, Harrison points out that twentieth-century readings of voluntarism derive from the how the term was defined in the late nineteenth century.[49] Application of such a meaning to the seventeenth century is simply anachronistic.

The voluntarism hypothesis itself requires constant revision with each practical application, and has little explanatory power. Such a theory may actually be misleading. For example, Force concludes that God as *Pantokrator* is of primary importance to Newton's theology and science. Force notes therefore that Newton's views on miracles retains God as "both a generally provident celestial artificer and a specially provident Lord God fully capable of *both* directly interposing his will into created nature and of doing so in a way which contravenes, or 'violates,' created natural law."[50] However, Force then applies a procrustean redefinition of voluntarism in Newton's case in order to take account of concerns directly related to how voluntarism might have been purportedly applied by Newton. In doing this, what is left unexplored is the question of the agency by which God, as lord of all, directly interposes God's will on created nature. Demonstrably, it is this very question of how to describe the agency by which the divine omnipotent lordship is exercised which is a major concern for Newton. Focusing on a weak theory misses the key issue and also misses a hitherto unacknowledged common thread in all of Newton's scientific, theological and alchemical work.

46. Brooke and Cantor, *Reconstructing Nature*, 21.

47. Harrison, "Voluntarism and Early Modern Science," 2. Henry, "'Pray Do Not Ascribe That Notion to Me': God and Newton's Gravity," 135.

48. Oakley, "Christian Theology and the Newtonian Science: The Rise of the Concept of the Laws of Nature."

49. Harrison, "Was Newton a Voluntarist?," 63.

50. Force, "Natural Law, Miracles and Newtonian Science," 91.

Aether and Spirit

Directly contrary to Foster, Newton believed in the spiritual (ensouled) nature of all creation and animate and inanimate objects within it. Newton's understanding of the spiritual nature of created things will be shown to be intimately related to his notion of the aether as the substance of spirit, that all things have a spiritual element, and that it is this spiritual element which is the stuff of life. This concept becomes for Newton the bridge for bringing an understanding of divine agency in humans to a similar understanding of divine agency in the world. While Newton never settled in his own mind that aether was the substance of spirit, he remained convinced that God's direct relationship to the world was the agency which both inspired and sustained it. Newton seems to have tried to find in nature evidence to support theological notions about spirit relationship to nature and people. In this, as in his biblical and scientific studies, he was seeking a common key that would enable full understanding.[51] Newton speculated at various times during his life that the substance of spirit was what he supposed as aether. Rather than being purely metaphysical his proposed that this medium was much more rarefied than air and was the substance of vacuum. He wrote about in a letter concerning the properties of light to Oldenburg,

> [T]here is an aethereall Medium much of the same constitution with air, but far rarer, subtiler and more strongly Elastic. Of the existence of this Medium the motion of a Pendulum in a glasse exhausted of Air almost as quickly as in the open Air, is no inconsiderable argument. But it is not to be supposed, that this Medium is one uniforme matter, but compounded partly of the maine flegmatic body of aether partly of other various aethereall Spirits, much after the manner that Air is compounded of the flegmatic body of Air intermixt with various vapours and exhalations. For the Electric and Magnetic effluvia and gravitating principle seem to argue such variety. Perhaps the whole frame of Nature may be nothing but various Contextures of some certaine aethereall Spirits or vapours condens'd as it were by precipitation, much after the manner that vapours are condensed into water or exhalations into grosser Substances, though not so easily condensible; and after condensation wrought into various formes, at first by the immediate hand of the Creator, and ever since by the power of Nature, which by vertue of the command Increase and Multiply, became a complete Imitator of the copies

---

51. In relation to Celestial motion, see Newton, *The Principia*, 4; In relation to prophecy, see Newton, *The Prophecies of Daniel and the Apocalypse*, 15–22.

sett her by the Protoplast. Thus perhaps may all things be originated from aether.

At least the electric effluvia seem to instruct us, that there is something of an aethereall Nature condens'd in bodies.[52]

What is interesting is that he proposes various aetheral substances, gravity, electricity, magnetism, the ability to reproduce, as well as life itself. Newton also suggests that aether and hence spirit might actually be measurable by the damping of the pendulum bob in vacuo. Aether would have a viscosity 10,000 to 100,000 times less than Mercury. Newton even designed an experiment using a dampened pendulum to measure this viscosity. He never published, presumably as the results were inconsistent. They would have depended on what we would now identify as energy losses through the stiffness of the string and friction in the pivot.[53] Newton's letter predates notions of heat energy or even its precursor concept of phlogiston. What is surprising is Newton's notion that the stuff of life is actually aetheral spirits. It is in this aetheral way that God who is spirit is in direct contact with the world.

Newton was accustomed to considering elements of the material world in terms of their having spirit and soul. This was not unusual among practitioners of alchemy, as Iliffe describes, "The alchemical tradition . . . held all nature to be alive, seemed to promise answers to question concerning fermentation, heat, and putrefaction, as well as the growth of animals, plants and minerals."[54] Newton as alchemist wrote "*Anima est medium inter spiritum et corpus utrique adhaerens*" (Soul is the medium in which the spirit and body together adhere.).[55] Newton's alchemy parallels his physics and parallels his theology. Rather than one influencing the other, his consistent early speculation in each area may point to his ideas on Spirit, aether and their relationship to matter being his synthesis of these notions. This is a contentious assertion as Westfall and Cohen indicate:

> The issue of Newton and alchemy is actively debated, frequently with passion, among Newtonian scholars. Most of those who have studied the large volume of alchemical manuscripts at length are convinced that his chemical experimentation needs to be understood as alchemical experimentation and that, in more general terms, alchemy needs to be seen as an important dimension of Newton's intellectual life. Other Newtonian scholars insist, with at least equal vehemence, that the alchemical

52. Newton, *Correspondence*, 1:364 (Dec. 7, 1675).
53. Ibid.; Newton, *De Aere et Aethere*, 39. Newton, *De Gravitatione*, 35.
54. Iliffe, *Newton: A Very Short Introduction*, 54.
55. Newton, "Index Chemicus." *The Chymistry of Isaac Newton*, 6.

papers are nothing but reading notes and that the arcane spirit of alchemy was antithetical to the Newtonian enterprise in science.... [T]hose who have pursued this subject argue that Newton saw in alchemy a form of natural philosophy that mitigated the harsh outlines of the mechanistic philosophy he had found in Déscartes and Gassendi; part of the attraction of alchemy was a philosophy that asserted the existence of nonmaterial agents in nature and the primacy of spirit over matter in the universe.[56]

Nonetheless, there is a commonality in Newton's terminology of the spirit which supports the first of these three options—that it is an important dimension of Newton's intellectual life. If a selection of Newton's thoughts on aetheral spirits are placed together without distinguishing between the writings as theological, scientific or alchemical a consistent picture appears. In explaining his concept of aether for his own benefit, Newton wrote,

> Thus this Earth resembles a great animal or rather inanimate vegetable, draws in ethereal breath for its daily refreshment & vital ferment & transpires again with gross exhalations, And according to the condition of all other things living ought to have its times of beginning youth old age & perishing. This is the subtle spirit which searches the most hidden recesses of all grosser matter which enters their smallest pores & divides them more subtly than any other material power what ever. (not after the way of common menstruums by rending them violently asunder etc.) this is Natures universal agent, her secret fire, the only ferment & principle of all vegetation. The material soul of all matter which being constantly inspired from above pervades & concretes with it into one form & then if incited by a gentle heat actuates & enlivens it.[57]

The parallel between nature and humans is explicit in this passage. Clearly his view of the relation of spirit to humans followed the shape of the Augustine's metaphysical soul. Newton further asserts that humans, as well as matter and living creatures and the whole of nature are inspired. At this point it is possible to argue that the case for inspiration as an influence in Newton's development of science is upheld. However, gaining a more detailed picture of the structure of Newton's thought helps with understanding why he did not make clear statements publicly outside his own notes and letters to friends. This has little to do with the strength of his firm convictions which shaped his ongoing influence and more to do with exposing his

---

56. Cohen and Westfall, *Newton*, 300.
57. Newton, "Of Natures Obvious Laws & Processes in Vegetation," 3v.

theological heterodoxy to public scrutiny and opening himself to criticism. It is also necessary to address an opinion that Newton's aetheralism was merely an immature fancy left behind by in his mature years.[58] Before arguing against this opinion, it is useful to show that Newton's inspiring aetheralism specifically has aspects that mark it as Augustinian *ekstasis* inspiration.

The influence of Augustine is readily demonstrated. Amongst notes dealing with chemical reactions, his early concepts of motion, Cartesian vortices, and comets, Newton titles pages with the terms of the Augustinian revision of the Aristotelian sensorium: Vision, colour, sound, odours, touching, memory and imagination.[59] Augustinian detail is evident in Newton's earliest work on optics, his Trinity notebook indicating that the Augustinian revision of Aristotelian metaphysical anatomy is in his thoughts from a very early period:

> The nature of things is more securely & naturally deduced for their operations out upon another than upon our senses. And when by the former experiments we have found the nature of bodys, by the latter we may more clearly find the nature of our senses. But so not clearly distinguish how far an act of sensation proceeds from the soul and how far from the body.[60]

Newton also made notes regarding these attributes in relation to matter and their spatial extension with the soul in the body and brain.[61] An essential part of the Aristotelian sensorium revised by Augustine is the ability of the spirit to cause physical action. This ability is in addition to the functioning of sight, hearing, touch, taste and smell in which Newton was deeply interested in describing and experimentally testing.[62] Newton applies this Augustinian anthropology to his aetheralism of spirit in physical action in another letter to Oldenburg. Newton supposes, "there is such a Spirit, that is, that the Animall Spirits are neither like the liquor, vapour or Gas of Spirit of Wine, but of an aethereall Nature, Subtile enough to pervade the Animal juices as freely as the Electric or perhaps Magnetic effluvia do

---

58. An opinion held by Koryré, Guerlac, and the Halls. Hall and Hall, "Newton and the Theory of Matter," 79.

59. Ibid., 35–39, 41–42; Iliffe, 32. This notebook attributes a short Latin quote explicitly to this source. Newton, "*Questiones Quædam Philosophiæ*," 41.

60. Newton, "*Questiones Quædam Philosophiæ*," Add. Ms. 3996.

61. Ibid., 33–34.

62. Newton, "Questiones Quædam Philosophiæ 16—," Additional Ms. 3975, pp. 1–22, 15.

glass."⁶³ Newton then describes how this might work roughly two centuries prior to the discovery of bioelectric currents.

> Thus may therefore the Soul by determining this aethereall Animal Spirit or Wind into this or that Nerve, perhaps with as much ease as Air is moved in open Spaces, cause all the motions we see in Animals: for the making which motions Strong, it is not necessary, that we should suppose the aether within the muscle very much condenst or rarified by this means, but onely that it's Spring is so very great, that a little alteration of its density shall cause a great alteration in the pressure. And what is said of Muscular motion may be applyed to the motion of the heart, onely with this difference, that the Spirit is not sent thither as into other muscles, but continually generated there by the fermentation of the Juices, with which its flesh is replenished, and as it is generated, let out by starts into the braine through some convenient ductus to perform those motions in other muscles by inspiration which it did in the heart by its generation. For I see not, why the ferment in the heart may not raise as Subtile a Spirit out of it's juices to cause these motions, as rubbing does out of a glasse to cause electric attraction.⁶⁴

This spiritual nature of motion expressed in a "scientific" writing is mirrored in his alchemical writing as well as being linked to the theological notion of God being able to be Lord of all, i.e. God is able to affect all things in all places at all times. In these writings Newton is explicitly concerned only God's capacity to act and does not mention anything about God's will which is so central to Foster's voluntarism.

> The ability of spirit to be able to move to all the recesses of matter is an attribute of the spiritual. It becomes an essential element in ensuring that the divine command can be exercised at all times and places. This not only related to the laws of nature but to the processes of life and vegetation. That vegetation is the sole effect of a latent spirit & that this spirit is the same in all things only discriminated by its degrees of maturity & the rude matter . . . How things conserve their species & how a tree might be conserved & nourished . . . Of protoplasts that nature can only nourish, not form them, That is God's mechanism.⁶⁵

---

63. Newton, *The Correspondence of Isaac Newton*, 1:368.
64. Ibid., 1:369.
65. Newton, "Of Natures obvious laws & processes in vegetation." *The Chymistry of Isaac Newton*, 1r. Vegetation has the sense of growth in Newton's use.

Normal behavior of nature is inspired by God's command. How God is able to reach into every part of everything helps to shape Newton's theory of matter and reflects what he describes as the abilities of the spirit. God must be able to act on all parts of matter, therefore God must be able to move through matter which implies that matter is not made of solid particles but must be something else entirely.

Matter in a Nutshell: Permeable to the Spirit

Was Newton's aetheralism something he abandoned in favor of a more mature description of mechanism at a distance, as the Halls suggest?[66] No, in spite of the Halls concluding that Newton was (in his mature years at least) an upholder of a "mathematical and vacuist"[67] and atomist,[68] "second-order mechanism and had rejected the aether."[69] However, this assertion is denied by Newton himself in a letter to Oldenburg after Newton developed his "venerated" optics, gravitation and mechanics.[70] The mature Newton held this same opinion some three decades later, as reported by David Gregory in late 1705.

> Sir Isaac Newton was with me and told me that he had put 7 pages of addenda to his book of light and colour in this new latin edition of it . . . His doubt was whether he should put the last quaere thus. What the space that is empty of body is filled with. The plain truth is, that he believes God to be omnipresent in the literal sense; And that we are sensible of Objects when their Images are brought home within the brain, so God must be sensible of every thing, being intimately present with every thing: for he supposes that as God is present in space where there is no body, he is present in space where a body is also present. But if this way of proposing this his notion be too bold, he thinks of doing it thus. What cause did the Ancients assign of Gravity. He believes that they reckoned God the Cause of it, nothing else.[71]

---

66. Hall and Hall, "Newton and the Theory of Matter."
67. Ibid., 78.
68. Ibid.
69. Ibid.
70. Newton, *Correspondence*, 1:364 (Dec. 7, 1675).
71. Thackray, "Matter in a Nut-Shell," 89; McGuire and Rattansi, "Newton and the 'Pipes of Pan,'" 104–5; Hall and Hall, "Newton and the Theory of Matter," 87.

The point is this: God as spirit fills not only vacuum but also all solid objects. This includes atoms which Newton did not consider solid particles of matter. The Halls read his understanding of the building blocks of matter as solid particles in their reading of this passage. They write: "Notice again the consistent Newtonian dual linkage between . . . God and absence of matter."[72] Rather than being antithetic to Newton's views on divine omnipresence God ability to move freely through "solid" matter remained important to him into his mature years.

However, as indicated earlier in his later letter on optics to Oldenburg, Newton explains that solid particles freely allow aether or spirit to pass. So why do the Halls persist? The Halls ask "If we may be forgiven here for reiterating . . . then he was indeed introducing a great new idea, analogous to and preparing the way for that of the field in nineteenth-century physics."[73] However, Newton could not have consciously prepared for the nineteenth- and twentieth-century theories. The incomprehension the Halls express was not one which Newton was likely to have suffered in understanding and communicating his own ideas as he used them.

Another example of problematic presentism relates to current use of the term aether. In Newton's time aether was not a fixed concept. Dobbs draws an important distinction between that of Newton's notion of a vital or spiritual aether as opposed to that of Déscartes' mechanical aether.[74] The mechanical nature of Déscartes' aether acting in vortices by direct contact is explicitly rejected at length by Newton in his *de Gavitatione*.[75] Also, aether considered as a mechanical fluid might have been used to explain the world atheistically without the "[C]ounsel of an intelligent Agent."[76]

Solid yet permeable, Newton's spiritual theory of matter seems strange. It was, however, a particularly important concept in the development of the significant science of chemistry.

Thackray suggests is that Newton's inspiration theory of matter continued to be importantly influential in chemistry. Thackray notes, "Lavoisier's declaration about the nature of matter was entirely in accord with the then-prevailing assumptions of Newtonian matter theory" and that "the theologically advanced and rationalistically inclined Priestley took the opportunity to abolish matter-spirit dualism by calmly abandoning matter."[77] As chem-

---

72. Hall and Hall, "Newton and the Theory of Matter," 87.
73. Ibid., 79.
74. Dobbs, "Newton's Alchemy and His Theory of Matter."
75. Newton, *De Gravitatione*.
76. Newton, "Questions from the Optics," 177.
77. Ibid., 94.

istry developed in the following century it followed this spiritualized theory of matter rather than the mechanistic materialist views which developed in physics. In chemistry as well as physics, Foster's main assertion is proven wrong: science advanced while maintaining a variety of spiritual understandings of matter. Not only were they spiritual they also had in common the idea that by some agency a perfect-being God somehow maintained matter and the universe by direct inspiration.

Cosmic Strings (after a Fashion): Newton's Gravity

Given Newton's statement that the material soul of all matter was "being constantly inspired from above pervades & concretes with it into one form & then if incited by a gentle heat actuates & enlivens it,"[78] then how did Newton think gravitation worked in such a God inspired universe? While it is possible to glean some detail from Newton's writings (published during his own lifetime, or expressed in his correspondence) additional detail can be drawn from his unpublished writings.

The Influence of Divine Perfections and Augustine in *de Gravitatione*

Newton insisted that the normal behavior of nature is inspired by agency of God alone. This also follows what was understood about the outworking of perfect-being theology in nature, and shows further Augustinian influence. Newton, corresponding with Thomas Burnet on his Theories in his *Sacred Theory of the Earth*, echoed Augustine's *de Genesi* in understanding that the six days in Genesis are not simply literal but rather are for our benefit.[79] Newton also parallels the sentiments of Augustine indicating the accommodation of the Holy Spirit to and overcoming of the limitations of the author.[80] In both cases because it is divinely inspired, it is accommodation without falsehood. Not only does Newton regularly utilize Augustinian inspiration, he was well acquainted with Augustine's key work describing inspiration, *de Genesi*, utilising it a number of times in *de Gravitatione* to

---

78. Newton, "Of Natures Obvious Laws & Processes in Vegetation," 3v.

79. Force and Popkin, *The Books of Nature and Scripture*, xvii; Mandelbrote, "A Duty of the Greatest Moment," 149–78; Newton, *The Correspondence of Isaac Newton*, 2: letters 246–47

80. Augustine, *de Genesi ad Litteram* 5.8.23.

criticize Déscartes.⁸¹ Dempsey sees Newton's comments on the nature of space and matter in *de Gravitatione* as being in opposition to Cartesian physics and dualism.⁸²

> Space is an affection of a being just as a being. No being exists or can exist which is not related to space in some way. God is everywhere, created minds are somewhere, and body is in the space that it occupies; and whatever is neither everywhere nor anywhere does not exist. . . . And hence it follows that space is an emanative effect of the first existing being, for if any being whatsoever is posited, space is posited. And the same may be asserted of duration: for certainly both are affections or attributes of a being according to which the quantity of any thing's existence is individuated to the degree that the size of its presence and persistence is specified.⁸³

This echoes then commonly held aspects of divine perfection. Newton wrote that nothing can exist which is not related to God. Augustine wrote: "Therefore, the providence of God rules and administers the whole creation, both natures and wills: natures in order to give them existence."⁸⁴ Newton argues all beings exist in space and are related to God who gives them motion and duration. According to Augustine, "The whole corporeal creation, therefore, does not receive extrinsic assistance from any corporeal source. . . . But intrinsically it is helped by an incorporeal force, since it is God who makes it possible for it to exist, For from Him and through Him and in Him are all things."⁸⁵

Through the later part of *de Gravitatione* Newton utilises Augustinian concepts. Countering Déscartes' notion that aetherial vortices provide motion, Newton wrote: "The parts of space are motionless. If they moved, it would have to be said either that the motion of each part is a translation from the vicinity of other contiguous parts."⁸⁶ Newton argues only objects can have motion. This echoes Augustine's notion of space in relation to its creator. Augustine claims God attributes motion not to space but to bod-

---

81. Whereas Dempsey cites this as undated, Dobbs redates this to before the *Principia*. Osler, "The New Newtonian Scholarship and the Fate of the Scientific Revolution," 8.

82. Dempsey, "Written in the Flesh: Isaac Newton on the Mind-Body Relation," 420–41.

83. Newton, *De Gravitatione*, 25–26.

84. Augustine, *de Genesi ad Litteram* 8.23.44.

85. Ibid., 8.25.46.

86. Newton, *De Gravitatione*, 25.

ies corporeal or spiritual as well as assigning their extension, location and separation.

> Therefore, almighty God, who sustains all things and is always the same in His immutable eternity, truth, and will, without moving through time or space, moves His spiritual creation through time, and also moves His material creation through time and space. Consequently, by this motion He rules the beings which by His interior action He has made, ruling them extrinsically both by wills subject to Himself which He moves through time, and by bodies subject to Himself and to those wills, moving these bodies through time and space—in that time and space whose reason-principle is life in God beyond time and space.[87]

Newton parallels these notions of divine motivated motion and imposed extension. "That for the existence of these beings it is not necessary that we suppose some unintelligible substance to exist in which as subject there may be an inherent substantial form; extension and an act of the divine will are enough."[88] Newton posits a parallel between God's capacity to act in the world and the human ability to act on objects in the world around them. The nature of the connection which allows the mind to act was for Newton spiritual.

> Thus I have deduced a description of this corporeal nature from our faculty of moving our bodies, so that all the difficulties of the conception may at length be reduced to that; and further, so that God may appear (to our innermost consciousness) to have created the world solely by the act of will, just as we move our bodies by an act of will alone; and, moreover, so that I might show that the analogy between the divine faculties and our own may be shown to be greater than has formerly been perceived by philosophers.[89]

Newton further draws a more overt Augustinian link making the connection between the human body and the mind interchangeable with the connection between a human body and divine will. The mind cannot sense or act without connection to the sensorium.

> But should anyone object that bodies not united to minds cannot directly arouse perceptions in minds, and that since there

---

87. Augustine, *de Genesi ad Litteram* 8.26.48.
88. Newton, *De Gravitatione*, 29.
89. Ibid., 30.

> are bodies not united to minds, it follows that this power is not essential to them, it should be noticed that there is no suggestion here of an actual union, but only of a capacity of bodies by which they are capable of such a union through the forces of nature. From the fact that the parts of the brain, especially the more subtle ones to which the mind is united, are in a continual flux, new ones succeeding those which fly away, it is manifest that that capacity is in all bodies. And whether you consider divine action or corporeal nature, to remove this is no less than to remove that other faculty by which bodies are enabled to transfer mutual actions from one to another, that is, to reduce body into empty space.[90]

Given how Newton uses this Augustinian anthropology of spirit and body in his other writings, this is unsurprising. Also, here as elsewhere Newton posits God's connection to the world in similar Augustinian terms.

> And so some may perhaps prefer to posit a soul of the world created by God, upon which he imposes the law that definite spaces are endowed with corporeal properties, rather than to believe that this function is directly discharged by God, To be sure, the world should not be called the creature of that soul but of God alone, who creates it by constituting the soul of such a nature that the world necessarily emanates [from it]. But I do not see why God himself does not directly inform space with bodies, so long as we distinguish between the formal reason of bodies and the act of divine will. For it is contradictory that it (body) should be the act of willing or anything.[91]

Dempsey merely sees this as Newton departing from Déscartes. It is more than this. In making his point, Newton drew heavily on Augustine as a source for grounding his understanding of the nature of matter and the relationship between the soul and body.

His commitment to this particular description becomes clear as will be shown in analysis of Newton's general scholium to *The Principia* and Queries 28 and 31 of *Optics* and his supporting arguments. There is further evidence of the importance of this description to Newton in the correspondence between Clarke and Leibniz. In each case, not only is Newton familiar with the Augustinian description of inspiration, but he applies the notion of a soul to nature and draws a parallel between divine agency acting in one and in the other.

90. Ibid., 34.
91. Ibid., 30–31.

Newton elsewhere argues that God, instead of working through a purported soul of the world, would act directly in such a soul's place. "[S]ome may perhaps prefer to suppose that God imposes on the soul of the world, created by him . . . But I do not see why God himself does not directly inform space."[92] Gravity is inspired directly by agency of God's spirit according to Newton. This occurs either by the possible spiritual substance aether or simply directly. The Spirit of God would stand permanently in the place that would have been occupied by a soul of nature. It is stood aside—*ekstasis*. God, according to Newton, inspires the action of gravity into matter. Such inspiration of inanimate matter is maintained perfectly and it consistently parallels how Newton understood divine inspiration of humans worked by the Holy Spirit in relation to the writing of Scripture.

The Sensorium of God

Newton realized there was an additional key task as he developed his natural philosophy. It was how to describe divine agency in nature as essential. Newton rightly saw a risk that his own philosophy would make God marginal to physical working descriptions of the universe.[93] Westfall notes Newton's concern that mechanistic explanations of the universe held the risk of atheism. Newton, like More, wished to reinstall the Spirit in the continuing operation of nature.[94] The difficult question was how to describe God's connection to the world without falling into what he saw as errors associated with the philosophy of his time. The first error, described in detail later in the discussion of the Leibniz-Clarke correspondence, was to avoid stating that gravitation and other properties were inherent to matter, which made matter independent of God and thus God unnecessary. The second was to avoid a "mechanistic" Cartesian-like description of a world that would also operate effectively without Divine input and even without God.

In reflecting upon his natural philosophy for public consumption, Newton sought to frame his explanation of the working of nature in such a way that the hand of the creator is always present. His General Scholium—summary to the second edition of *The Principia*—emerged as the result of a long period of reflection. This was a lengthy process. Newton was aged 24 in 1666, the "miraculous" year he developed the calculus, the laws of

92. Newton, *de Gravitatione et Aequipondio*, in McGuire and Rattansi, "Newton and the 'Pipes of Pan,'" 107.

93. Alexander, *The Leibniz-Clarke Correspondence*, xiv; see also Newton, *Correspondence*, 1: letter 146; 5: letter 918.

94. Westfall, *The Life of Isaac Newton*, 30.

gravitation, mechanics and optics. Publication of the *Principia* did not occur until 1687 and only after Hooke and Halley urged him to publish. The intervening two decades was Newton's period of greatest concentration on alchemy and theology. It was a further 26 years before the Second edition of the *Principia* included the summary now known as the General Scholium in 1713. By this time, Newton had become aware of the a-theological trends developing in the application of his Natural philosophy. Thus, the General Scholium to the *Principia* and the similar comments in his *Optics* were part of his attempt to counter a trend exclude God from natural philosophy.

Stewart also expresses a common scholarly dissatisfaction with historical scholarship's explanation of the genesis of the General Scholium or Newton's intentions.[95] Stewart notes that Newton was a master of hermeneutics and existing theories do not fully explain the nuances in Newton's thought.

General Scholium

The primary attributes of God that Newton wished to convey in relation to Natural philosophy include God's dominion over all the universe and God's ability to act in the world at every time and place, while remaining separate from the world. In the General Scholium, after commenting on the beauty and perfection of the arrangement of the solar system, Newton moves the discussion onto a theological plane asserting that God, the *Pantocrator*, has control of the universe as something other than God's own being. The implication is that God must be involved in all action in all the world.

> This Being governs all things, not as the soul of the world, but as Lord over all; and on account of his dominion he is wont to be called "Lord God" παντοκρατωρ, or "Universal Ruler"; for "God" is a relative word and has a respect to servants; and Deity is the dominion of God, not over his own body, as those imagine who fancy God to be the soul of the world, but over servants.[96]

Why does he emphasize that God is not the soul of the world immediately after claiming God's dominion over this non-God universe? Why is this important? The clues continue in the remainder of the general scholium. God must not be considered part of the world, but rather the lord over it. "The Supreme God is a Being, eternal, infinite, absolutely perfect,

---

95. Stewart, "Seeing through the Scholium: Religion and Reading Newton in the Eighteenth Century," 139.

96. Newton, *The Principia*, III.

but a being, however perfect, without dominion, cannot be said to be Lord God."[97] For Newton perfection is an attribute of God's being as the consequence of God's omnipotence. This omnipotence must extend to God also being in control of God's own existence. Tautologically, God must have dominion over all things including God's self in order to be lord of all. God exercises dominion over the world. God's inherent perfection is expressed by perfection in action. How God exercises this dominion and how divine action occurs in the world is something that Newton is careful in the scholium to leave as unclear—even unknowable. The next section of the general scholium suggests however that he did have an idea, an analogy, in mind.

> Every man, so far as he is a thing that has perception, is one and the same man during his whole life, in all and each of his organs of sense. God is the same God, always and everywhere. He is omnipresent not virtually only but also substantially; for virtue cannot subsist without substance. In him are all things contained and moved, yet neither affects the other; God suffers nothing from the motion of bodies, bodies find no resistance from the omnipresence of God. It is allowed by all that the Supreme God exists necessarily, and by the same necessity he exists always and everywhere. Whence also he is all similar, all eye, all ear, all brain, all arm, all power to perceive, to understand, and to act; but in a manner not at all human, in a manner not at all corporeal, in a manner utterly unknown to us.[98]

The confusing point about this paragraph concerns Newton's coupling of God's omnipresence with organs of perception. A footnote cites Anaxagoras in this paragraph. It is significant that Newton's footnote refers to Anaxagoras. Anaxagorus' sensorium; sight, hearing, memory, touch, perception and understanding and the ability to act, is revised by Aristotle and later adapted by Tertullian and Augustine. Newton's rider, "Not at all corporeal" echoes in the same words Augustine's critique of Tertullian's ravings in *de Genesi ad Litteram*. This incorporates Augustinian anthropology and metaphysics. However, the question remains, why use any description of human anatomy when talking about the relationship between God and the world? The answer to this is no different to what has been shown in his unpublished writings. Newton is arguing by analogy to claim that the agency of God's Spirit in the world was similar to the agency of God's Spirit in humans, i.e. the Holy Spirit acts while standing the soul to one side of the sensorium. The final clause of the scholium provides a clearer statement

97. Ibid.
98. Ibid.

of this anatomy from this period of history when the electrical nature of nerves was unknown.

> And now we might add something concerning a certain most subtle spirit which pervades and lies hid in all gross bodies, by the force and action of which spirit the particles of bodies attract one another at near distances and cohere, if contiguous; and electric bodies operate to greater distances, as well repelling as attracting the neighbouring corpuscles; and light is emitted, reflected, refracted, inflected, and heats bodies; and all sensation is excited, and the members, of animal bodies move at the command of the will, namely, by the vibrations of this spirit, mutually propagated along the solid filaments of the nerves, from the outward organs of sense to the brain and from the brain into the muscles. But these are things that cannot be explained in few words; nor are we furnished with that sufficiency of experiments which is required to an accurate determination and demonstration of the laws by which this electric and elastic spirit operates.[99]

"Electric" refers to the phenomena of what is now known as static electricity, sustained electric current being unknown in his time. Ironically, Newton's development of glass technology to improve lenses for his optical work led to a wide spread interest in static electricity.[100] Iliffe notes while Newton was president of the Royal Society towards the end of his life, he conducted experiments demonstrating the phenomena of electroluminescence (static electricity.) Iliffe points to the inclusion of reference to "electric" in *The Prinicpia* as a late revision dating from this period. It is the concept of animal motion, the ability to act, to move a limb which, in Augustinian anatomy, is the action of the spirit upon the sensorium which allows the human will to act on the world around the person. Newton's analogy is between the human spirit which pervades a human body, and the spirit who pervades all matter. Newton has, however, already clearly stated that God is not the soul of the universe. So how can he draw this analogy that has God working in the world in the same way spirit senses and acts through the human body? The key is the spiritual nature of the soul. If the human soul is stood to one side, then God—who is not part of the person— may act through the person's spirit in this way. Such a standing to one side is *ekstasis* and is the agency of God's action in Augustine's description of in-

---

99. Ibid. Iliffe, *Newton: A Very Short Introduction*, 116.

100. Fara, *Newton: The Making of Genius*, 93–94. Iliffe, *Newton: A Very Short Introduction*, 116.

spiration. The analogy only works if God, as it were, stands aside the "soul of the universe" and senses and acts through space and time. Then God would work directly in the universe in every place without being part of creation.

Understood in this way, the odd juxtaposition of the ideas in the general scholium begins to make sense. However, why did Newton, who was clear on so many other issues, not make this analogy clearer? Perhaps Newton here anticipated the controversy that followed the clearer statement made later in one of his queries in his *Optics*.

Optics—Queries 28, 31

The relevant comment from Optics, Query 28, in which Newton refers to infinite space as the sensorium of God is as follows:

> And these things being rightly dispatched, does it not appear from phenomena that there is a Being, incorporeal, living, intelligent, omnipresent, who in infinite space, as it were in his sensory, sees the things themselves intimately and thoroughly perceives them, and comprehends them wholly by their immediate presence to himself, of which things the images only carried through the organs of sense into our little sensoriums are there seen and beheld by that which in us perceives and thinks? [101]

Originally, Newton worded this Query 28 more explicitly, "Is not infinite Space the Sensorium of a Being incorporeal . . . " Newton, as discussed earlier, was accustomed to considering elements of the material world in terms of their having spirit and soul through which God worked sustaining life, growth, and the order of nature.[102] "That the soul of matter was inspired from above."[103] Having decided he had been too bold in his statement he attempted to recall all copies of this edition and reword it as quoted above.[104] In spite of this, he failed to recall them all and paste the new page into every copy. An unaltered copy fell into Leibniz's hands and retentive memory[105] resulting in scathing criticism.

A usual mistake made in discussing the sensorium comment is to understand it applying only to the five senses. Alexander, in his introduction

---

101. Newton, "Questions from the Optics," 135–79, query 28.
102. Explained earlier in section 3.3.
103. Newton, "Of Natures obvious laws & processes in vegetation," 3v.
104. Westfall, *The Life of Isaac Newton*, 259–60; Lit., *Annon Spatium Universum, Sensorium est Entis*.
105. Fara, *Newton: The Making of Genius*, 110.

to the Leibniz-Clarke correspondence, makes this typical mistake in seeing sensorium as Newton accepting the representative theory of perception in its extreme form.[106] As has been demonstrated, Newton consistently included the additional terms of memory, understanding and the ability to act in the overall sensorium. What Alexander sees as an "odd expression" is actually the Aristotelian concept revised by Augustine.

The following comment from queries 23 & 24 more clearly demonstrates Newton's assumption of this anthropology in which he likens the motive power for animal motion to the same media as that of the senses. This is in keeping with his notes and letters previously discussed. "Is not vision performed chiefly by the vibration of this medium, excited in the bottom of the eye by the rays of light and propagated through the solid, pellucid, and uniform capillamenta of the optic nerves into the place of sensation?"[107] "Is not animal motion performed by the vibrations of this medium, excited in the brain by the power of the will and propagated from thence through the solid, pellucid, and uniform capillamenta of the nerves?"[108]

These vibrations could only be understood to be electricity after the 1850's.

Newton sounds a theological caution in query 31 of *Optics*, explicitly stating:

> a powerful ever-living Agent, who being in all Places, is more able by his Will to move the Bodies within his boundless uniform Sensorium, and thereby to form and reform the Parts of the Universe, than we are by our Will to move the Parts of our own Bodies. And yet we are not to consider the World as the Body of God, or the several Parts thereof, as the Parts of God. He is an uniform Being, void of Organs, Members or Parts, and they are his Creatures subordinate to him, and subservient to his Will; and he is no more the Soul of them, than the Soul of Man is the Soul of the Species of Things carried through the Organs of Sense into the place of its Sensation, where it perceives them by means of its immediate Presence, without the Intervention of any third thing. The Organs of Sense are not for enabling the Soul to perceive the Species of Things in its Sensorium, but only for conveying them thither; and God has no need of such Organs, he being every where present to the Things themselves. [109]

---

106. Alexander, *The Leibniz-Clarke Correspondence*, xvi.
107. Newton, "Questions from the Optics," query 23.
108. Ibid., query 24.
109. Ibid., 177–78.

The implications of both of this and the sensorium statement have been difficult for scholars to interpret together. Dempsey comes close, noting the similarities between Newton's description of the relationship between mind and body in humans and Newton's assertion of a parallel relationship between God and the world in *de Gravitatione*: "I might show that the analogy between the divine faculties and our own may be shown to be greater than has formally been perceived by philosophers."[110] Dempsey describes Newton's ideas as some "interesting sort of mind-body monism"[111] but is unable to trace its roots other than identifying a Hebrew source and Socinian influences. What Dempsey identifies as Newton's mind-body monism is actually Augustinian metaphysical anatomy.

Newton needed an interpretive framework which would allow him to both assert that infinite space can be the sensorium of God but at the same time that God is not part of space. The Augustinian description of inspiration provides this framework. That is, in the act of inspiration, the person's will is displaced and the sensorium of the person becomes, albeit temporarily, God's to use, while at the same time God is not the person. Thus Newton's understanding of God's work in the universe parallels the received Augustinian description of the agency of the Holy Spirit during the inspiration of humans.

(Newton) Clarke-Leibniz Correspondence

Some of the theological implications of Newton's assumptions are worked out in the correspondence between Newton's long term rival, Leibniz, and Newton's protégé, Clarke, as they sought to win the opinion of the Princess Caroline. Clarke and Newton failed to persuade her. It is widely recognized that the opinions Clarke defended and expressed were Newton's.

This correspondence consists of a series of five letters from Leibniz and replies from Clarke exchanged during 1715-6. The exchanges terminated with the death of Leibniz in 1716 and were later published by Clarke. Clarke's reason for publication and Leibniz's for his active participation were to enable the further pursuit of truth as well as to defend natural religion and the dominion of God from atheistic elements in natural philosophy. After a polite commendation of the late Leibniz in his dedication of the

---

110. Newton, *De Gravitatione*, 30.

111. Dempsey, "Written in the Flesh: Isaac Newton on the Mind-Body Relation," 420.

assembled letters, Clarke asserts, "Christianity presupposes the truth of natural religion . . . God's continual Government of the world."[112]

Alexander summaries the main themes of the correspondence as 1) the appropriateness of Newton's claiming space as the sensorium of God and 2) the protection of divine perfection particularly in the principle of sufficient reason.[113] Despite mid-twentieth century Newton scholarship's claim that divine voluntarism is vital to Newton's thinking, the term is not used once in the correspondence. The ideas related to perfect-being theology are never in dispute, however the application of their differing physics is. Neither protagonist disagrees with the principle of sufficient reason acknowledging: that nothing happens without a cause; God must always have a motive for acting, and; God must always act for the best.[114] Leibniz is particularly concerned that Newton's philosophy "derogates from God's perfection"[115] requiring perpetual miracles in contrast to Leibniz's "pre-established harmony"[116] in nature. This debate is about the appropriateness of utilising the analogy of sensorium and how this might support or take away from divine perfection. It asks, what is the nature of the relationship between body and soul? Is it appropriate to use this as an analogy to God? It also asks what perception is, and where and how does it occur in space and time. Between two intellectual heavyweights whose mutual hostility led them to agree on very little, it is that which they assume in common and on which they tacitly agree which speaks more loudly, giving weight to the influence of the Augustinian inspiration shaped understanding of divine agency in humans to having been used to shape the understanding of divine agency in the world.

Leibniz: Metaphysics and Monadology

Leibniz's interaction with Clarke and Newton dealt with the question: By what agency does God act in and through the world? As Clayton has indicated, Leibniz's writings are diverse and extensive. It is necessary for this case study to confine discussion to the assumptions Leibniz makes in relation to divine action, revelation and Scripture, and his understanding of the makeup of the natural order.

---

112. Alexander, *The Leibniz-Clarke Correspondence*, 6.
113. Ibid., xiii.
114. Ibid., xxii.
115. Ibid., 39, Leibniz, letter 4.
116. Ibid., 85, Leibniz, letter 5.

This discussion draws on the works of his last years *Discourse on Metaphysics* and *Monadology* in addition to the content of his correspondence with Clarke, which dates from the same period. Leibniz's starting point in his early work *Discourse on Metaphysics* is a variant of the sufficient cause argument making God the Perfect architect. "God is a perfect being. Consequently power and knowledge do admit of perfection, and in so far as they pertain to God they have no limits."[117] Further, "I think that one acts imperfectly if he acts with less perfection than he is capable of. To show that an architect could have done better is to find fault with his work."[118] This leads logically to the case he makes for Scripture as one of the two books of God's revelation. God acts in or with a human agency to produce Scripture in such a way that God being the perfect architect cannot work anything less than perfection. Thus, Scripture as God's work must be perfect in revelation, truth, beauty and simplicity. The laws of God in nature must also be similarly perfect.

> He who acts perfectly is . . . like a good householder who employs his property in such a way that there shall be nothing uncultivated or sterile; like a clever machinist who makes his production in the least difficult way possible and like an intelligent author who encloses the most of reality in the least possible compass.[119]

Leibniz concludes by dismissing those who would argue against the final causes of the laws of physics, thus:

> That position seemed to go the length of discarding final causes entirely as though God proposed no end and no good in his activity, or as if good were not to be the object of his will. I hold on the contrary that it is just in this that the principle of all existences and of the laws of nature must be sought, hence God always proposes the best and most perfect.[120]

In *Metaphysics*, Leibniz discussed the workings of the human mind and soul within its senses, in a manner consistent with the Augustine's anatomy of the mind. The soul, the essence of the person, is attached to the person through the senses and memory and thereby able to act.[121] Leibniz's focus here, however, is not inspiration. He is not concerned about *ekstasis*

117. Leibniz, "Metaphysics," 1.
118. Ibid., 3.
119. Ibid., 5.
120. Ibid., 19.
121. Ibid., 33.

but rather with the ultimate eternal perfection of the soul or spirit within the person by God's action. Leibniz came to weave the notion of God as perfect architect with his own description of the nature of matter in the debate with Clarke on Newton's theories. He describes these ideas in *Monadology* written just prior to the correspondence with Clarke. All substances, atoms, particles, spirits are made from unique simple monads.[122] He considered matter and even souls to be made from these enduring monads.[123] Eternal life, he argued, consists in the perfection of the spirit by God but not in its endurance of being which it already possessed as an attribute of its substance.[124] Yet, all substances are unique and eternal; no two monads could be identical. Leibniz's argument is that if they were identical then they would be perfectly identical in size, shape, nature and place and position in time and space. They would therefore be the same monad. God relates to each monad as an individual, which has its existence from the point of creation, even though its attributes may and do change.

For Leibniz, the laws of God in nature are perfect because they emerge as the result of God's perfect design and forethought. Whereas Leibniz attributes the consistency of God's laws in nature to the properties of matter, Newton saw them resulting from the direct action of God. Ironically, as with the calculus, it is Leibniz's description of the laws describing gravity and light as inherent to matter that is commonly attributed to Newton as a supporting plank of Newtonian physics and later Newtonian Deism. In contrast, Newton indicates that their perfection arises by being continually sustained by agency of the divine perfect-being inspiring the universe. Leibniz's physics, while post-Cartesian, were not compatible with Newton's.

What Explanation Best Supports Perfection?

Although the place of natural religion and the Lordship of God are never in dispute between the parties, Leibniz nevertheless supposes they may be jeopardized by Newtonian natural philosophy. Clarke argues in turn that Leibniz has not understood Newton's attempt to preserve both of these essentials in his philosophy. By using sensorium language Newton was, Clarke argued, preserving the lordship and providence of God at every point in time and space.

Leibniz's second letter alludes to the common foe in contemporary materialism, which in imitation of the atomism of Democritus and

---

122. Leibniz, "Monadology," 3–5.
123. Ibid., 18–19, 61–65.
124. Ibid., 19–30.

Epicurus and Hobbes described the world without admitting any spiritual component. He opens his first letter lamenting the decay of natural religion because some people held the human soul to be material and God corporeal.[125] In this, Leibniz echoes Augustine's criticism of Tertullian's adaptation of corporeality of spirit from the same classical atomistic writers who were influencing the secularists in Leibniz's day.[126] The implication here is that anyone who asserts that God is somehow corporeal has moved from the faith, and encourages others to do the same. Newton, he claims, does this by making space an organ of God and further claims that Newton makes God less than a perfect creator by having a God who "wants to wind up his watch from time to time : otherwise it would cease to move."[127]

In reply, Clarke states that the clock-maker illustration "tends to exclude providence and God's government in reality out of the world."[128] After agreeing with Leibniz about the theological implications of the materialists, he states that the term "sensorium" is only a similitude. In humans, the sensorium and brain are the place in which our picture of the world is formed. This is not needed by God as the author and creator of all who already has this picture without using the sensorium. It is providence which is important; God must have the ability and the means to be sovereign at all times and places. The sensorium notion provides for God's direct involvement in the world. Otherwise, God can be excluded when describing creation. If God is can be excluded then obviously God is not at all times and in all places creation's Lord.

Leibniz states in his last letter: "the harmony or correspondence between the soul and the body is not a perpetual miracle; but the effect or consequence of an original miracle, worked at the creation of things; as all natural things are."[129] Leibniz accused Newton of saying that God works in nature by a perpetual miracle. This is not a compliment, for in his principle of sufficient reason there must be some reason or principle in any divine choice. "God's perfection requires that all his actions should be agreeable to his wisdom; and that it may not be said of him, that he has acted without reason; or even that he has prefer'd a weaker reason before a stronger."[130] However, the down side, as Clarke explains, is that if the world runs according to an established pre-planned design then this becomes fate rather

---

125. Alexander, *The Leibniz-Clarke Correspondence*, 17, 20, Leibniz, letter 2.
126. Augustine, *de Genesi ad Litteram* 7.20—7.21.26–29.
127. Alexander, *The Leibniz-Clarke Correspondence*, 11, Leibniz, letter 1.
128. Ibid., 14, Clarke reply 1.
129. Ibid., Leibniz, letter 5.89.
130. Ibid., Leibniz, letter 5.19.

than providence.[131] For Newton, God's perfection as creator meant perfect total lordship. Leibniz could not see how this ultra-micromanagement of the universe could avoid making God part of it and constantly affected by it.

The irony is that Leibniz, while holding to contingency, was criticized by Clarke for risking the elimination of freedom in describing the world as perfectly pre-planned and ordered. Newton, while defending the providence of God, was in essence accused by Leibniz of making the future of both the world and God subject to contingent change at every point. While the perfection of God's action was never in doubt, there was no clarity on how to define that perfection of action. Contrasting Leibniz's ideal design against Newton's ideal dominion highlights the difficulty already discussed in deciding which of the perfections of God's freedom should be given priority.

Body and Soul, World and Spirit?

Leibniz's opposition to Newton is twofold. Firstly, as shown, he rejects the need for continuous divine action. Secondly, he is never convinced that the sensorium language can work as a mere analogy but must imply that God is somehow corporeal and part of the created order. Significantly, Leibniz initially does not question the analogy as an analogy. He is fully aware of the Augustinian sensorium language and the classical underpinning for this interrelation of the soul and spirit—even to the point that Clarke accuses him of blithely following later scholastic explanations of this anatomy.[132] Further, Leibniz does not dispute, or even question, how it is that God acts within a human person in inspiration. He assumes along with Clarke that the human soul acts within the sensorium.

Leibniz ends his second letter with a dual conclusion that continuous miracles are a hypothesis reduced *ad absurdum* and that if the action of God in the world becomes natural then God is part of the world which, under Newton's model, would make God the soul of the world. Newton is reluctant to state openly the sensorium language in terms any stronger than similitude, as this would imply that the world actually had a soul. There are difficulties with taking a stronger approach, not the least of which would have been that Newton's unorthodox theology would become publicly apparent. Newton claims, only as a "similitude," that the "displacement" of the world's soul by God would be as if the Spirit of God permanently displaces the "soul of the world" for all created time. There is a parallel between this replacement of "soul of the world" in creation and certain non-orthodox

---

131. Ibid., Leibniz, letter 4.1–2; and Clarke, reply 4.1–2.
132. Ibid., Clarke, reply 4.

notions that the divine mind of Christ removes the need of a natural mind in the human Jesus. Newton personally would not have seen such a parallel as a particular problem with his own idiosyncratic Christology, but he would have been aware that it would constitute a problem for others. That there was a heterodox theological implication was not lost on Leibniz. In his second letter, Leibniz immediately implies that Newton's theories are in keeping with the theology of Socinus who held that Christ's human nature was replaced by the divine logos after the resurrection.[133] Leibniz's accusation was not one that Newton, or indeed Clarke, could easily counter. In fact, Clarke never attempts to counter this accusation. Leibniz implied correctly, in comparison with the canons of traditional orthodoxy, that Newton had an imperfect conception of God. Leibniz's concern about Newtonian natural philosophy having ties to theological heterodoxy continued to be justified throughout the next century.

In addition to questioning non-Trinitarian theological implications, Leibniz did not think that the analogy, as Newton and Clarke suggested, was appropriate in describing the relationship of God to the world or as God's immediate presence to the world.[134] While Newton and Clarke emphatically denied that God was part of the created order, Leibniz was not alone in appreciating Newton's link as a difficulty which has led a number of scholars to interpret Newton's position as an early form of panentheism.[135] In his second and third letters Leibniz, expounds what he believes sensorium language implies. Firstly, it explicitly means that the sensoria are actual organs making Clarke's claims of mere analogy incomprehensible in the act of sensation. Secondly, Leibniz argued that it is impossible to conceive of sensation and perception occurring without affecting the soul. That is, God could not be present in the world in this way without both destroying divine impassibility or subtracting from God's perfection. The trouble that Leibniz saw was that, even in using this soul/body link as only an analogy, it made God out to be what God could not be.

The question of where and how perception occurs in relation to the sensorium and soul is why Leibniz and Clarke debate space and time in their correspondence. Leibniz accuses Newton and Clarke of being too simplistic in their definition of how perception occurs. They believe it is the presence of the metaphysical soul in a sensorium that allows perception of the world to occur.[136] In Leibniz's thought this simple description, "this vulgar notion,"

133. Alexander, *The Leibniz-Clarke Correspondence*, Leibniz. letter 2.10.
134. Ibid., Clarke, reply 2.12.
135. Brooke, *Science and Religion: Some Historical Perspectives*, 143–44.
136. Alexander, *The Leibniz-Clarke Correspondence*, Clarke, reply 2.12.

that has reality represented in mind through perception conveyed by the organ of sense is inadequate even for the Cartesians he considered to be wrong.[137] In contrast, Clarke and Newton do not attempt to describe how awareness occurs, but only that it does. Leibniz was not pleased with this claim of *hypotheses non fingo*[138] (I frame no hypotheses). Leibniz argued that correctly to understand the interaction of soul and body requires making note of how and where perception could occur. What Leibniz rejected as inadequate was Clarke's simple assertion that the presence of the soul is what allows it to perceive what passes in the brain and senses. Newton, however, was deeply interested in describing and experimentally testing how sensation worked. Newton's experimentation included testing his own body to extremes. In one such extreme example, he described self-experimentation using a bodkin to deform the back of his eye. Leibniz's own revision of the Augustinian description of human soul/body anatomy made Newton's claim of analogy less likely to be valid in his view. This is especially so in the light of Leibniz's particular understanding of physics. According to Leibniz's scheme, all properties and interactions must be the property of something, whether the property is gravitational attraction, position or perception and recollection. These properties could not exist in what he considered a chimerical vacuum. Leibniz agreed with Aristotle's *Physics* in that he did not believe in either vacuum or infinity. "If space is a property or attribute it must be the property of some substance."[139] Gravity, if not a property of matter, had to be a class of monad as yet unknown. Leibniz believed Newton had made gravity a chimera. He was also particularly unhappy that Newton makes of gravity what he sees as almost a spiritual phenomenon. As has been demonstrated, Newton did consider it to be spiritual.

Part of the difficulty in the debate is that Clarke, following Newton, is very careful to merely assert that the universe as observed works in the manner described, never venturing to answer the cause. For instance, Newton is very careful to say that gravitational attraction depends on the mass and distance between object, but he does not to attempt to describe how it works—either as force at a distance or some other connection. He does however rule out Déscartes' theory of vortices as incompatible with the observed motion of the planets. Similarly, he and Clarke are careful not to say how God orders and perceives the world or even how the sense organs transmit sensations or how the thought moves limbs. Rather they say only that these happen.

---

137. Ibid., Leibniz, letter 5.84.

138 Newton, "*Questiones Quædam Philosophiæ* 16—," 63–67. Newton, "Of Colours," 15.

139. Alexander, *The Leibniz-Clarke Correspondence*, Leibniz, letter 4.8.

Newton adopted conventions for his physics that did not sit well with Leibniz including having identical bodies moving in vacuum, infinite space and absolute time and space. This was incompatible with Leibniz's understanding of the world, matter, substances and how he understood perception to work. Leibniz's notions now seem unusual. In his own time, they had the force and benefit of great antiquity as well as the virtue of being divorced from classical authors like Democritus and Epicurus, both of whom he—with Clarke—saw as contributing to atheistic natural philosophy.

In their discourse, the Augustinian *ekstasis* description of inspiration was not in dispute only its application in a novel way. Newton's tentative suggestion was that the inspiring action of God ensured that the Laws of God in nature remain as consistent and perfect in the universe as they are observed to be in practice. Newton's suggestion that God writes God's laws in nature by a process of taking the place as it were of the soul of the world was, by Leibniz's reasoning, incomprehensible.

God's Law Revealed by Augustinian *Ekstasis* Inspiration of Infinite Space

When criticized by Leibniz for anthropomorphising God, Newton claimed he used the term "Space as the sensorium of God" as a metaphor. It is interesting that Leibniz believed that Newton was claiming more than a metaphor, which is what Newton had in mind given the earlier discussion of Newton's unpublished theory of matter. Leibniz did not misunderstand the shape of the illustration. Rather, he thought that Newton was unavoidably making God part of creation. He criticized Newton's use rather than the analogy itself. Newton's reference to a form of anthropology that had classical and patristic roots was not controversial. Tying God to nature in a similar fashion was. What is relevant here is that the sensorium language in Newton's reflects the same Augustinian anatomy of the mind and senses which forms the foundation for the *ekstasis* description of inspiration. Newton was implying publically and explicitly in private that the mind of God stands in relation to space as sensorium in a manner similar to the way the human mind was supposed to stand in relation to the sensorium of the body, the five senses, motion and memory. Moreover, the analogy becomes complete only if the mind of God by the Holy Spirit stands aside a fictive mind of the world in the same way that the Holy Spirit was understood to stand aside the human mind in inspiration.[140] Newton drew a link between Augustinian anthropology and creation. Harrison has noted that many of Newton's

---

140. Harrison, *The Fall of Man and the Foundations of Science*, 239.

contemporaries were particularly interested in Augustinian anthropology, particularly as it relates to the fall of humanity. Harrison has stated that Newton was agnostic "about the fallen state of human nature." However, by playing such a significant role in Newton's philosophy, the *ecstasis* description of inspiration ties Newton closely to Augustine's anthropology, albeit a different aspect than that of the Fall. In this respect, Harrison's argument for the significant influence of Augustinian anthropology in the foundations of science can be demonstrated to be also present in Newton.

What lay behind Newton's use of this Augustinian analogy? He desired to ensure that God is at all times and everywhere in interaction with the creation and that God has the ability to act anywhere and at any time in the world by the Spirit. In addition, the interaction of bodies in space should at all times follow the perfect ordered patterns he had been able to describe. There was confusion about Newton's meaning, even amongst his colleagues. Newton refused, however, to accept that properties like gravitational attraction remained an inherent property of matter. Newton's combination of preserving Divine providence at all points of space and time, in having God "as it were, the mind of the universe," makes the laws of God in nature to be simultaneously a reflection of the mind of God and also the result of God's on-going free choice to be the Lord of all. God could and did choose to imbue atoms of matter with additional properties as "the soul of matter was inspired from above."[141] These additional properties as in *ekstasis* inspiration remain imbued only while the divine action continues. He disclaimed the notion that gravitation was something inherent to matter which allowed it to act at a distance.[142] In a series of letters to Bentley, he insisted emphatically: "[D]o not attribute the idea that gravity is inherent to matter to me."[143] He was similarly critical of Cotes making a similar assertion in his preface to the Second edition of the *Principia*.[144] He was even cautious about admitting that gravity was a force that could act at a distance.

The dominant theory Newton was trying to replace was that of Déscartes. According to Déscartes, the mechanism of hypothetical vortices in the aether by direct mechanical contact with the object caused attraction. Newton saw this problem in Cartesian philosophy: if the world operated as a simple mechanism of any description, then God need not have any part in the explanation of the world. That is, apart from Descartes' theory failing to predict observed data. The difficulty he had with Cotes' and Bentley's

---

141. Newton, "Of Natures obvious laws & processes in vegetation," 3v.
142. Newton, *The Correspondence of Isaac Newton*, 3:233–56.
143. Ibid., 5:397.
144. Ibid.

popularisation of his natural philosophy was that by making Gravitation or the action of light inherent properties of matter, they thereby reduced nature to a different form of mechanism than that of Déscartes—but a mechanism all the same which could be explained without God. Newton saw concerns about the atheistic implications of the presentation of his own ideas when others remained oblivious.

Newton linked to the sensorium the ability to move limbs by action of the spirit as a way to describe how the omnipresent God can and does act on the world at all points whilst not becoming part of it. Further, as highlighted in the interaction with Leibniz through Clarke and in other correspondence, Newton understood that God continuously acts to imbue matter and space with whatever properties, forces or other mysterious mechanism which make matter and space obey God's law. If, however, God were to cease this action then gravity and the propagation of light and material existence would cease. Creation for Newton is directly and continuously sustained by the hand of God. Newton's opinion differs significantly from the "set and forget" Deistic "Newtonian" clockwork universe. Such an idea was abhorrent to Newton. If the universe worked as a clock, it was because God moved each hand and each cog at every moment. Ironically, the "Newtonian" clockwork universe conceived as created flawlessly and then running with perfection without further involvement from the creator is more consistent with Leibniz's perfect architect argument, than with Newton's ideas.

The link between the Augustinian anatomy of the mind and Newton's reference to space being the sensorium of God is more than a simple metaphor, if what Newton meant is as follows: There are no such things as the laws of nature, for example gravity, as inherent properties of matter in the natural world. In fact, the very ordering of nature comes about by God acting through nature as sensorium, giving both the appearance and substance of perfect order. That is, God acting through space as sensorium acts in exactly the same fashion as it was believed that God acted through the inspired writer of Scripture. The laws of God were, Newton understood, written perfectly in the world by the agency of permanent *ekstasis* inspiration of infinite space. There is no doubt for Newton that they are God's laws in nature and not the laws of inherent to nature as they came to be understood in the next century.[145] The Augustinian description of inspiration thus is not merely an influence on the development of modern science, but is foundational to Newton's philosophy. Thus Newton drew together the notion of

---

145. Harrison, "Newtonian Science, Miracles, and the Laws of Nature"; Oakley, "Christian Theology and the Newtonian Science: The Rise of the Concept of the Laws of Nature."

the two books, and perfect-being theology to apply the description of divine agency in humans contained in inspiration to divine agency in the world.

After Newton

Newton's description of divine agency in nature as an underpinning for natural philosophy continued to receive complex and varied support in the years following his death. Newton and Leibniz disagreed only on how best to apply the notion of the two books of God's revelation and perfect-being theology to nature. In particular perfect-being theology expressed in the principle of sufficient reason continued to be highly influential: that nothing happens without a cause; God must always have a motive for acting, and; God must always act for the best. The conflict between Newton's perfect dominion and Leibniz's perfect order foreshadowed future difficulties that developed into the nineteenth century. These difficulties will be discussed in relation to Darwin and Huxley in the next chapter. One of the important theological issues during the latter part of the eighteenth century was how to best affirm the inherent harmony of nature as the best of all possible worlds as the reflection of divine perfection. The non-christological nature of Newtonian promotion of Newton's ideas allowed Protestant and deist theologians and natural philosophers to find common expression against those who either rejected God or all trappings of traditional theology or God's relevance to the natural order. Conversely, as worried both Newton and Leibniz, Newton's ideas were also adapted to radical theology and atheism.

Although Newton's theories were scientifically sound, they carried with them the implications and stigma of a heterodox theology. The ongoing influence of Newton's spiritual theory of matter on Chemistry has already been mentioned. Priestley as a pioneer of chemistry developed Newton's ideas oddly—a spiritual materialism—as he states of matter that whatever

> other powers matter may be possessed of it has not the property that has been called *impenetrability* or *solidity* From the manner of expressing our ideas we cannot speak of powers or properties but as powers and properties of some *thing* or *substance* though we know nothing at all of that thing or substance besides the powers that we ascribe to it and therefore when the powers are supposed to be withdrawn all idea of substance necessarily vanishes with them.[146]

---

146. Priestley, *Disquisitions Relating to Matter and Spirit*, 32.

That is matter is nothing but an expression of a spirit who imposes properties of repulsion, attraction and the appearance of solidity. That is all matter is constantly upheld by God. "If they say that on my hypothesis there is no such thing as matter and that every thing is spirit I have no objection provided they make as great a difference in spirits as they have hitherto made in substances."[147] While content to argue that all matter is spirit, God was for Priestley different to matter. Priestly was strangely a materialist who rejected the notion of a soul in that no material being could have a part which was part of the great indivisible spirit—God.[148] Priestley's God was a generic divine perfect-being essentially unlike the Christian God. Priestley highlights a problem in using generic perfect-being theology in understanding divine agency. Priestley began with Newton's revision of aether based on non-christologically adapted Augustinian understandings of spirit but moved away from both Augustine and Christianity. He was perceived to be so influential that he was put on trial for heresy and his laboratory destroyed by a mob in protest at his ideas.

In the other developing disciplines of natural philosophy Newton's theories were also adapted in markedly different ways. Those following Newton often did not fully appreciate or rejected what was at stake for Newton in explaining nature in a way that made God's omnipotence and omnipresence essential. Voltaire, for example, was a strong proponent of Newtonian philosophy in relation to mechanics and physics, but he rejected the idea of providence which was the driving thought behind Newton's sensorium analogy.[149] The important thing Newton wanted to avoid was any mechanistic model which would remove the need for God and thereby lead to the atheism he despised. However, it is precisely such a notion that all matter inherently attracts all other matter by the force of gravity which is commonly attributed to him. A parallel problem has occurred in the last century as Einstein regretted naming his theory "relativity," which has commonly been misapprehended as supporting relativism—the view that how things appear depends on your point of view. In truth, the point Einstein makes is the exact opposite—that light and motion obey exactly the same rules, exactly the same, irrespective of any point of view. He should have used the term "theory of invariance."[150]

---

147. Ibid., 33.

148. Ibid., 73ff.

149. Chadwick, *The Secularization of the European Mind in the Nineteenth Century*, 239–40.

150. Yu, "The Principle of Relativity as a Conceptual Tool in Theology," 183.

Newtonian aetheral theories depending on Augustinian inspiration were also developed. For example, Cheyne and Hartley continued to explore aether models. Hartley proposed a physiological theory for the functioning of nerves which became "enormously influential in France"[151] and remained influential well into the nineteenth century according to Huxley.[152] Cantor and Hodge indicate "Newtonian" aether theories proliferated and diversified widely.[153] Newton's protégé Desaguliers, the Royal Society's experimental demonstrator, often openly and freely stated opinions that Newton had cautiously abbreviated and hesitantly inserted into his published work.[154] General acceptance of these notions among the general public was enhanced by Desaguliers' vigorous and entrepreneurial promotion. Fourier a mathematician and forerunner of sociology considered Newton a semi-divine authority, posing a version of Newton's aether as the harmonizing fluid which bound people together in society.[155] Newton's suggestion of the spiritual nature by which the mind controls the body was also widely repeated even as far as the other side of the world. "All sensation is excited, and the members of animal bodies move at the command of the will, namely, by the vibrations of this spirit, mutually propagated along the solid filaments of the nerves, from the outward organs of sense to the brain."[156] One repetition at the far end of the earth was in a Tasmanian popular almanac, which also repeated for popular consumption a summary of observations about the natural world also including Charles Darwin's observations on plants and the geological sinking of mountains.[157]

There were also aetherial notions propagated in theories developed as an orthodox counter to Newton's theological influence. For example Hutchinson's alternative *Principia* viewed the cosmos as "a giant machine powered by an aetherial fluid that took three forms—light, fire and spirit, corresponding respectively to God, Jesus and the Holy Ghost."[158] Hutchin-

---

151. Fara, *Newton: The Making of Genius*, 87–89.

152. Huxley, "On Sensation and the Unity of Structure of Sensiferous Organs," 292–3.

153. Cantor and Hodge, "Introduction," in *Conceptions of Ether: Studies in the History of Ether Theories 1740–1900*.

154. Fara, *Newton: The Making of Genius*, 89–97.

155. Ibid., 189–91

156. Lovett, *Philosophical Essays*, 37. This explanation was commonly repeated. Martin, *A New and Comprehensive System of the Newtonian Philosophy, Astronomy and Geography*, 11; *The British Palladium*, 17 (a popular almanac); Hutton, *A Philosphical and Mathematical Dictionary*, 2:106.

157. *Hobart Town Magazine*, 60.

158. Fara, *Newton: The Making of Genius*, 107.

son's theory became popular among high Anglicans in the late eighteenth century.

Obscuring of Newton's Anti-Trinitarian Thought in Scriptural Studies

In the years following Newton's death the influence of his ideas and suspicions about his unorthodox theology both grew. Suspicion was fuelled in part by apologists excusing any hint of unorthodoxy in Newton's published works. Mandlebrote reports that some of these attempts were successful. For example, the early nineteenth-century president of Magdalen College denied unorthodox views in Newton's often published and translated *Two Notable Corruptions of Scripture*, as "hearsay testimony."[159] Snobelen describes as "Nicodemite" strategies," the way Newton and the Newtonians downplayed the implied theological implications as they propagated his ideas.[160] Nevertheless Newton's heterodoxy was known and suspicion countered.[161]

It is ironic that the treatment of those espousing Newtonian views differed markedly from the latter public memory of the man. Whiston was dismissed and Clarke's protégés were passed over for ecclesiastical preferment for publicly stating theological views they shared with Newton.[162] In spite of this, Newton later became more and more lauded as the shining example of scientific and moral genius. By the 1830's William Whewell, who coined the word scientist, was called on to protect Newton's reputation in the face of persistent (and well-founded) rumours about his character and his less-than-fashionable rational interest in alchemy. Brewster continued to the point of bowdlerising Newton's work. This defence Brewster claimed was to protect the cause of religion from injury.[163] Chadwick describes this protection of religion and the issue of where to place ultimate authority, as being one of the urgent questions in academic thought through the century

---

159. Mandlebrote, "Eighteenth-Century Reaction's to Newton's Anti-Trinitarianism," 110.

160. Snobelen, "Isaac Newton, Heretic: The Strategies of a Nicodemite," 384; Force and Popkin, eds., *The Books of Nature and Scripture*, xvi, 182; Force, "The God of Abraham and Isaac."

161. Wiles, *Archetypal Heresy*, 121.

162. Ibid., 139; Force, "The God of Abraham and Isaac," 181.

163. Yeo, "Genius, Method and Morality: Images of Newton in Britain 1760–1860," 271.

subsequent to Newton's life.[164] Should authority be placed in revelation, or Church leadership, or reason or in natural philosophy?

After Leibniz and Newton

As Newtonianism developed over the next century, the theological implications of Newton's ideas continued to be controversial. Nonetheless, Newton's ideas had utility, descriptive and predictive power and continued to be popularised, as did the scientific notions of Leibniz. In the most general terms, what emerges is a complex amalgam of Leibniz's theology and more accessible mathematics, mixed with Newton's theology and methodology for describing the world. There is little doubt that this did not always proceed smoothly and that there were problems of consistency which became apparent only as the details of their systems were explored more fully.

Reflecting on these problems in Leibniz's influence on theology, Barth offers an extended criticism of Leibniz's appeal to God as the guarantor of nature.[165] Barth points out the great logical conundrum of Leibniz's *Theodicy*—God is perfect but the world in pain and sin is somehow of an order of goodness less than absolute perfection. Leibniz's logic, Barth argues, is without fault. The problem, according to Barth, is not Leibniz's logic but the assumption of an idea bearing a close resemblance to God's self-revelation in Jesus Christ but independent of that revelation. Barth traces the development of Leibniz's thought in the work of his student Wolff who further perfects the Leibnizian ideal of a clockwork universe. Ultimately Barth concludes that Wolff and then Leibniz's mechanistic description of the universe depends on an assumption of the two books theory—but that that assumption is flawed. According to Wolff, then, nature must consistently follow laws in preference to requiring direct divine intervention. "God therefore maintains more good by permitting evil and wickedness than by not doing so. He can thus have no reason to render it impossible by means of miracles."[166] The next logical, step is to remove the requirement for the miraculous and indeed to argue that miracles are evidence of a lack in divinity. Barth asks why there would be any such need "when the perfection of the world was in any event guaranteed by the clockwork character affirmed by

---

164. Chadwick, *The Secularization of the European Mind in the Nineteenth Century*, 71.

165. CD III/1:388. Barth refers to Leibniz's *Essays de Theodicee* (1710); *Discours de metaphysique* (1686); and *de rerum originatione radicali* (1697).

166. CD III/1:395.

Leibniz himself?"[167] Huxley was to later use similar logic to argue that belief in miracles could be construed as sinful.[168]

Barth argues that there is a particular inevitability in the downward development of Deism. "Once we have boarded this train, we find that it is a non-stop express and we must accept the fact that sooner or later we shall reach the terminus. And we can take comfort in the principle of the school—if the application may be permitted—that imperfection is integral and even essential to creaturely perfection, and serves only to increase it."[169]

Nonetheless, the divine agency in the world drawn as a parallel to that in humans in the Augustinian description of inspiration is part of the continuing legacy of this debate. More specifically, the legacy is an abiding assumption that God's laws are perfectly written by God in the created order. It is the challenging of this perfection in the reading of the book of nature which is next explored.

Vestiges of Divine Perfection in Nature

In addition to this new understanding of divine agency in the world, widespread support continued for applying the notion of the two books of God's revelation to understanding God's laws in nature, for the notion of divine perfection and for inspiration. There had to be cause for everything. God must always have a motive for acting and act for the best. The world is the best of possible worlds. The laws of God in Nature were put in place by God alone without error just as Scripture was seen to be. From the seventeenth to nineteenth centuries the view of the perfection of God's action in the world and the high regard thought appropriate for Scripture held the ground in spite of significant challenges being raised including those which led to Bayly, Noble and McCaul writing against Quakers, deists and material rationalists.

Newton was employed by Bayly (1708), in his attack on the Quakers as he argued for perfect divine action as a precondition of true inspiration by the Holy Spirit. He claims infallible foreknowledge of historical events written under inspiration of those whose predictive abilities were, "without much or deep Skill in this art as a Flamsteed, a Newton or a Halley."[170]

Noble (1823) in answering the criticisms of the deists and other "scoffers" vigorously describes a doctrine of inspiration of the infallible Scriptures as the total inspiration of its writers. His is a divine dictation description

---

167. Ibid., 395–96.
168. Desmond, *Huxley: The Devil's Disciple*, 288.
169. CD III/1:404.
170. Bayly, *An Essay on Inspiration*, 117.

assuming divine displacement of the faculties of the inspired. The writers are "prevented from recording any material error."[171] "The Scripture, which is declared, on this ground to be, what no partially inspired composition can be, absolutely be infallible."[172] Noble cites Origen's three levels of meaning, but nevertheless followed Augustine's error free view of inspiration over Origen's claim that inconsistency forces the reader to consider higher truth. This standard text develops the *ekstasis* description of inspiration as infallible dictation.

Answering Fitzjames Stephens, a rationalist secretary for the commission on education in 1862, McCaul reiterated a traditionalist line espousing again that the Scriptures are authoritative by virtue of their infallible inspiration.[173]

Favouring Natural Theology as One Book of Revelation over the Other

However, during the seventeenth and into the eighteenth century there is evidence of a change in emphasis where more and more trust began to be placed on reading the book of nature as a surer source of divine revelation than a set of writings that may have suffered corruption through historical transmission. Over time more and more confidence was given to the perfection of God's laws in nature as scientific methodology developed and showed continuing promise and effectiveness. For many this confidence of the perfection of God's action in nature came to be held more highly than for Scripture as the eighteenth century unfolded. While an extreme example, the heart of Priestley's influential criticism of the established order included an attempt to revise the Scriptures themselves in the light of scientific discovery.[174] Nevertheless in spite the official suspicion levelled at Priestley, the disciplined methods used to interpret Scripture were reflected in the study of nature where the harmony in nature had become the evidence of perfect divine agency. This widely held confidence then dramatically dropped during the course of the nineteenth century. It is the demonstrable absences of what was assumed to be "perfection" which becomes the foil which Darwin and Huxley use to reject traditional Christian faith.

171. Noble, *The Plenary Inspiration of the Scriptures*, 11

172. Ibid. 53. McCaul, *Testimonies to the Divine Authority and Inspiration of the Holy Scriptures*, 5, 57, 114.

173. McCaul, *Testimonies to the Divine Authority and Inspiration of the Holy Scriptures*,

174. Knox, "Dephlogisticating the Bible," 179–84.

The change in emphasis from an unquestioning belief that divine agency perfectly wrote or inspired the laws of nature to serious doubt about and limiting of the perfection of such divine action can be clearly seen in the development of the scientific discipline of geology. Geology initially had an intimate and then later a more remote relationship with Christian theology. Gillespie[175] and Rupke[176] both give excellent overviews of the discipline's development tracing changes in the relationship between the two fields of study in terms of the diminishing place of divine providence. Initially the assumption that the world would reflect the actions of the divine perfect-being was held in common. The nature of divine providence was expected to be demonstrable in the field.

The following survey is an example in one influential discipline of broader trends in natural philosophy as it developed into science. It contains no new or controversial material, but is necessary to trace the development of widely accepted ideas into concepts which Darwin and Huxley felt compelled to reject. In doing so they questioned certainty about divine action.

The Vanishing Book of Revelation according to Geology

Geology developed during the late eighteenth through to the mid nineteenth centuries. In the early stages of its development, Geology as a science largely bypassed England, but not Scotland, until the early nineteenth century. The practical research of John William Smith laid the foundation for the science's later popularity in England. The first school of geology in England was set up at Cambridge by William Buckland and came to include Adam Sedgwick and William Whewell. In contrast to continental geology, the English school was inculcated in Anglican tradition so that natural religion was intimately interwoven with the substance of the science. This kind of natural philosophy was also known as "divine philosophy" by the end of the eighteenth century. "It is defined as that spark of knowledge of God which may be had by the light of nature and the consideration of created things; and thus can fairly be held to be divine in respect of its object, and natural in respect of its source of information."[177] It is significant that the relationship of the new science to theology was initially very close. "The difficulty as reflected in scientific literature appears to be one of religion in science rather that one of religion versus science."[178]

175. Gillespie, *Genesis and Geology*.
176. Rupke, *The Great Chain of History*.
177. C. C. J. Webb cited in Gillespie, *Genesis and Geology*, ix.
178. Ibid.

By the early nineteenth century however, a change occurs in the tone of how this was expressed. As Gillespie states: "The emotion and ardour of seventeenth-century religious feeling had disappeared out of the divine order of nature by the middle of the eighteenth-century, the providential interpretation of the origin and purpose of the physical universe had not then been challenged in its essentials."[179] However, by the mid-nineteenth century dealing with assumptions about God's work was not the primary interest for those involved in the new work of science. There is a discernable reduction in both biblical and theological description in published scientific work. For example in Lyell's *Principles of Geology*[180] (1830-3) the Bible barely rates a mention.

One idea that did not survive was the harmonizing the two books of God's revelation, specifically harmonizing geology with the book of Genesis. While a trend developed to take Genesis more figuratively, there remained a reluctance to abandon the idea that the truth of Scripture could be reconciled with scientific evidence. The second idea that was abandoned with great reluctance, only after Darwin, was the supposed Law of continuity. This idea traced back to Leibniz as an indication of divinely created perfection. That is to say, all species were preserved from the ark onward.[181] God acting as perfect creator would not have created anything not suited to its environment, and thus species once created must still exist. The alternative would be to admit that God did not act perfectly. The idea that the whole chain of being from creation to the present should be linked was still strongly held, so that any break in that would mean a collapse in providence and God's perfect action. Thus there was no place for extinctions. Upsettingly for this notion, extinctions were the very thing that Buckland and Cuvier's hyena caves demonstrated: Once there existed tropical species in temperate England and France.

Controversies and Narrowing of Ground for Assumptions

The development of Geology was marked by a series of sometimes bitter controversies in which these changes can be illustrated. The first between Neptunists and Vulcanists, for example, shows changes regarding assumptions of the literal truth of the text. The later debates between Huttonian uniformitarianism and Cuvier and Buckland it is possible to trace changes in the priority given to each of the two books.

179. Ibid., 19.
180. Lyell, *Principles of Geology*.
181. Rupke, *The Great Chain of History*, 169–71.

Neptunism proposed that most geological structures such as strata could be explained by deposition from sea water in keeping with Mosaic accounts of the Flood. It should be noted that while Neptunists accepted these preconceptions it did not lessen the intensity or curiosity or vigor of pressing their studies.[182] Such harmonisation became secondary in responding to the Vulcanists who explained strata as being formed by magmatic heating and uplift.

The place of the book of Scripture started to move behind that of the book of nature as Huttonian uniformitarianism developed. The text came to be treated more figuratively, making way for what was discerned to be a more clearly established law governing the development of creation. The Huttonians were not specifically linked to natural theology but the design argument had found wide acceptance among them as well in other groups.[183] That is perfect design in nature in the operation of its laws or the interaction of creatures in their environment pointed to the involvement of God's action as creator. For example, Hutton's popularist, Playfair, assumed explicitly that God had created the world, but had not left any indication regarding when He had.[184] It was up to the Scientist to deduce how it had happened through rigorous investigation.

Gillespie states that by 1820 not even the most liberal champion of scientific freedom, had questioned the relationship of natural history to natural theology: "Though the age of the earth had been greatly extended and the effort to connect Holy Writ directly to earth history was now felt to have been discreditable to all concerned."[185] Indeed, certain beliefs had not even been questioned, notably belief in Noah's flood.[186] Gillespie sees these geological arguments narrowing the thread of interpretation tying Genesis to geology. As the conflicting groups attempted to interpret the growing masses of data from the field it became harder to link their competing arguments with the inspired narrative of Scripture. Gillespie reviews the history of academic thought which moves from a literal six days, to six thousand year time spans, to divinely ordained cataclysms rather than only one flood. Cuvier's[187] (1796) and then Buckland's[188] (1819) work on fossils indicated

---

182. Gillespie, *Genesis and Geology*, 52–66.
183. Rupke, *The Great Chain of History*, 18.
184. Gillespie, *Genesis and Geology*, 76–77.
185. Ibid., 96.
186. Ibid., 90.
187. Cuvier, "Living and Fossil Elephants"; Cuvier, "Megatherium from South America."
188. Buckland, *Reliqiae Diluvianae*.

that there had been a series of catastrophes in the past, thereby countering Huttonian uniformitarianism.[189] Both considered that God's action was perfect whatever that action actually was. As creator God would have ordered life perfectly no matter what the change or length of time or the catastrophe.

It is, however, fossils that forced reconsideration of some largely unchanged assumptions regarding natural theology and the Law of continuity.[190]

Perfection implied lack of any need to change as well as being good, healthy, functional and appropriate. What developed was the conception of a providential God the proof of whose existence depended upon the immutability of different manifestations of life.[191] Townsend, a student of Smith, explicitly stated such an all-or-nothing approach to faith in 1815 in his very popular *Recording Events Subsequent to the Deluge*. "The divine legation of Christ . . . must stand or fall together. If the Mosaic account of the creation and of the deluge is true, and consequently the promises recorded by him well founded we may retain our hopes; but should the former be given up as false, we must renounce the latter."[192] Nevertheless, scientific analysis of fossils indicated massive changes in species in the ancient past. The process of making sense of this new information led to the beginnings of catastrophist controversy.[193] William Buckland, a member of the middle Anglican tradition was head of the first school of geology in Oxford argued that the perfection of harmonious nature is the proof of God's action. Buckland's *Reliquiae Diluvianae*[194] presented evidence of the extreme age of the earth and he suggested that the six days of Genesis be taken as figurative.[195] This is not out of keeping with historical usage as previously described in Augustine's *de Genesi*[196]. As might be expected, Buckland's suggestion generated significant controversy in the 1820s from biblical literalists.[197] There was no common reaction to Buckland's theories in relation to this presentation of the hyena cave bones. Rupke suggests that across theology, denomination and scientific discipline there was only a commonality in their trust in the

---

189. Rupke, *The Great Chain of History*, 31.
190. Ibid.
191. Gillespie, *Genesis and Geology*, 147.
192. Ibid., 94.
193. Rupke, *The Great Chain of History*, 31.
194. Buckland, *Reliqiae Diluvianae*.
195. Gillespie, *Genesis and Geology*, 106.
196. Augustine, *de Genesi ad Litteram* 1.1.1.
197. Rupke, *The Great Chain of History*, 42.

written word of Scripture as opposed to natural objects.[198] While Rupke's proposition has merit, Harrison's suggestion is more useful—that a similar method for reading God's two books was used. Most of the opposition seems to have arisen out of incredulity that there could be any inconsistency between the two forms of revealed truth.

The development of divine agency as described earlier highlights that the problem giving rise to the controversy lay in the parallel drawn between divine agency in the creation of the book of nature and divine agency in humans for the production of Scripture. This understanding had become axiomatic, both for the literalists and for those suggesting revision. No one in the period of the formation of geology seems to have seriously questioned that underlying description that gave rise to commonly held assumptions that God's action was perfect, implying immutability of species and no extinctions. Rather, what occurred was that a series of conservative, even minimalist revisions, were offered. These steadily reduced the grounds on which the old assumptions could stand so that later revisions become more radical, as with Huxley's "evangelical" fervor for agnosticism.

Buckland presented a path of conservative revision, offering a way of reconciliation seeking to give full credibility to geology in God's inspiration of nature by the hand of providence as well as divine inspiration in the text of Genesis.[199] However, there were still prominent members of faculty such as Edward Nares at Oxford who continued to argue for a biblical literalist position even after Buckland's fossil data had been widely circulated.[200] Old explanatory narratives persisted even in the face of new evidence while bedrock assumptions remain unexamined. This was a major issue in the nineteenth century and remains, it is argued, a problem in the contemporary dialogue between theology and science.

Beginnings of the End of Providence Assured by God's Perfect Action

Having briefly described some of the foremost changes to assumptions during the development of geology and palaeontology before Darwin, it is important now to outline some of the background to the academic circles into which Darwin gives birth to his theories.

The beginning of the demise of providence and God's inspiring of the laws of nature as a generally accepted assumptions coincides with the early

198. Ibid., 48.
199. Ibid., 209.
200. Ibid., 62.

part of Darwin's career and the later part of the Catastrophist controversy. This provides the academic environment in which Darwin commences his work on the *Beagle* and in the decades leading up to the publication of *Origin*. Lyell's landmark text *Principles of Geology* appeared just as Darwin was to leave for South America. Darwin took the first volume with him on the Beagle and received the second at Montevideo.[201] Up until the 1830s, the focus of relating Scripture to geology had largely moved from attempts to demonstrate how Geology and six-day creation could be harmonized to a focus on post-diluvial geology—whether this was a single flood or multiple catastrophes as suggested by Buckland. While Lyell by the publication of *Principles* may have moved from his own earlier scriptural harmonising position, it is clear that he remained convinced that his earlier assumptions, particularly scriptural infallibility ensured by divine inspiration, provided an essential element of the Christian faith.[202]

Agassiz and Lyell have a marked change in their use of the term perfect and how it is reflected in nature. Agassiz extensively researched fossils and argued for a progressive modification of species among fossils. Each modification he maintained was a divinely designed step in the plan of creation. Each change and step up was anatomically perfect. His writing clearly entertained an anthropocentric definition of "perfect," for the more fossils were like humans the more perfect they were described to be.[203] Lyell on the other hand continued to see perfection reflected in unchanging original perfect design even if some species became extinct.[204]

Sedgwick, Buckland's opposite number at Cambridge, along with Whewell laid down a challenge to the uniformitarians to demonstrate how there could be a progression in natural forms as the result of a uniform unvarying law.[205] The influential response came in Robert Chambers' *Vestiges of the Natural History of Creation*.[206] Sedgwick still had a high view of Scripture. The taking of the bible as fable and the teaching that humans were descended from apes, he regarded as eliminating the distinction between physical and moral.[207] In response to this, Chambers answered that it is "ours to admire the perfection of the plan not to criticise it."[208] Both men

201. Desmond and Moore, *Darwin*, 108.
202. Lyell, *Principles of Geology*, 50.
203. Rupke, *The Great Chain of History*, 160–61.
204. Ibid., 186.
205. Gillespie, *Genesis and Geology*, 149.
206. Chambers, "Vestiges of the Natural History of Creation."
207. Rupke, *The Great Chain of History*, 177.
208. Gillespie, *Genesis and Geology*, 158.

assumed there still was a divine plan the hand of a divine perfect-being and therefore perfect providence was at work in the changes in the history of the world and in life itself.

The description of divine agency in world began with a generic understanding of divine perfections and inspiration. There is no Christology needed to state that: there had to be cause for everything; God must always have a motive for acting and act for the best, and; the world is the best of possible worlds. The agency for the production of the laws of God in nature and scripture were explained without the need to describe in any detail who the law giver might be. What was deduced from Barth was that any generic description of divine agency would be predicted, in the face of contradictory evidence, to develop into an unresolvable choice between finding new ways to assert absolute truths or to abandon traditional ways of speaking of God. In the development of geology as one part of the investigation of the natural world in the eighteenth and nineteenth centuries such contradictory evidence appeared. New ways were found to express the figurative truth of Scripture. The usage of "perfect" changed considerably from Newton and Leibniz to Lyell and Agassiz. There was a progressive limiting of perfection: from ideal in all aspects or perfection ensuring agreement between Scripture and nature; to perfection in design and natural law; to perfection in the ultimate purpose of creaturely design; and, to no perfection in design as espoused by Darwin. The next chapter will trace the development of Darwin's and Huxley's rejection of perfection.

Rather than inspiration providing a mechanism for understanding divine agency in the world which implied creation's perfection, it came to be held that God's perfect design was considered evidence that proved divine inspiration. This development in the late eighteenth century is exemplified in Paley's use of the perfect adaptation of creatures to their environments as evidence of the hand of the perfect work of the creator. Paley's formulation is explicitly attacked by Darwin and is given as his confessed rationale for an earlier loss of faith.

This perfection sits uncomfortably with a number of key Christian theological concepts. In what way is the incarnation perfect, if God chooses to become subject to the limits of human finitude? This was a problem for the eighteenth-century deists as well as the nineteenth-century agnostic. How can God deal with sinful, imperfect humanity? How can the presence of sin and evil in the world be explained in a "perfect" world, where perfection is taken as the proof of Divine action?

The advent of Darwinism, as Gillespie rightly describes, is Paleyanism inverted. To suggest that species have adapted to fit their environment and so as a result are apparently suitable to their environment is the apparent reverse of the argument from design. While this does not exclude the work of a creator in using this method to bring about this adaptation of the creature, it runs contrary to the Paleyan notion of immutability.[209] It is to the development of Darwin rejection of divinely perfect action in nature and later Huxley's rejection of this and the existence soul to which it is now appropriate to turn.

---

209. Ibid., 219.

CHAPTER 4

Divine Agency Implying Perfection and the Existance of the Metaphysical Soul

Perfection as Precondition Challenged by Science

What the previous chapter established in the work of Newton is that a parallel was drawn between divine agency in humans as contained in *ekstasis* inspiration and divine agency in the world. This agency was understood to complement early eighteenth-century understandings of the notion of the two books and divine perfections as factors contributing to how the manner of God's direct action in the world came to be described. While there was contention over the use of the term "soul of world," the nature of divine agency in the world was considered spiritual and that this was linked in some way to the existence of the human soul. It has been shown in the Clarke/Leibniz correspondence that there were problems about how to speak of the perfections of God's freedom. Which freedom should have priority? As indicated in the second chapter Rogers, Clayton and Barth, while advocating different solutions to speaking of the divine perfections, all acknowledge that this difficulty is inherent in the early modern understanding of perfect-being theology. The understanding of divine agency in the world which developed was that of a generic perfecting-being god, which could be—but was not necessarily—identified with the God of Christian tradition. As stated earlier drawing on Barth, one could surmise that any generic description of divine agency would be predicted, in the face of contradictory evidence, to develop into an unresolvable choice between finding new ways to assert absolute truths or to abandon traditional ways of speaking of God. The apparent choice became between finding new ways to affirm God's perfections or turn to experientialism.

Having demonstrated the development of the generic understanding of divine agency in the world, exploration is required to show how this understanding develops into an unresolvable tension in the dialogue between theology and the newly professional discipline of science in the late nineteenth century. This is particularly relevant as both orthodox and deist came to consider that nature reflecting the divine perfect-being and the existence of the soul were preconditions of the Christian faith. In the nineteenth century, however, revision of the Christian faith seemed unavoidable when imperfections in the book of nature were discovered and the absolute veracity of inspired revelation was challenged.

It is useful to examine the reasons for the rejection of these implications by Charles Darwin and by promoters of his ideas such as Thomas Huxley. It is necessary to glean relevant details from the wealth of correspondence and publications of the debates immediately following the publication of Darwin's ideas and Huxley's promotion of them. They wrote in the late nineteenth century in a period of great intellectual complexity, in which many debates that commenced never reached a point of resolution. Contemporary perceptions of the period's intellectual debates often continue to be coloured by questionable myths and deliberate historical revisionism. While, there is a good deal to be examined, there has however, as Bowler indicates, been little analysis of theological developments in response to science in the generations following the 1860's.[1] This chapter will work with what Darwin and Huxley actually wrote to determine the extent of the influence of this understanding of divine agency in the world on their thinking.

Three influences will be demonstrated to be significant in the nineteenth-century erosion of the Augustinian underpinnings of this description. Firstly, Newton's magisterial influence remained as a continuing inducement for non-Trinitarian theology. Secondly, the reading of the book of nature to deeper layers, both figuratively and literally, by the new sciences such as geology, palaeontology, biology and botany showed that nature was not as "perfect" as anticipated. Thirdly, the post Newtonian and Cartesian philosophers moved well away from the Augustinian "soul in the sensorium" metaphysical anatomy.

---

1. Bowler, *Reconciling Science and Religion*, 9, 2–3.

Respectable Deism—Paley

Despite serious attempts to repatriate Newton, the influence of his identifiably heterodox theological position was often seen as a dangerous influence at a time when there was strenuous debate between the competing orthodoxies of Protestantism and Catholicism. By the early nineteenth century, in the opinion of the conservative establishment, unorthodox theological positions were associated with radicals, free thinkers and murderous revolutionaries.[2] This suspicion by people in the governing classes continued in spite of increasing numbers of students and lecturers at universities in Scotland, England and Europe actively exploring the implications of alternative theologies.[3] The French revolution and the rise of Napoleon cemented this suspicion of radical ideologies among the British establishment. There is a real tension in this period between searching for truth in science and political expediency. For example, research on evolutionary ideas and transmutation of species was discouraged in Darwin's university days to the point of censorship. Darwin's first public presentation describing polyps was at a Plinian meeting more noteworthy at the time for a talk given by the politically radical Browne who claimed that life was not supernatural or incorporeal implying no soul and no afterlife. Browne's talk was considered so inflammatory there was an unusual attempt by a member of the Plinian society to censor both Browne's talk and its notification from the minutes.[4] There is irony in that Charles Darwin—a member of the conservative landed gentry and establishment—came to a "radical" position regarding the laws of nature embracing transmutability of species. Desmond and Moore speak of Darwin's tortured life, his self-castigation as the "Devils Chaplain." He sat on his evolutionary theory for 20 years, and even when prodded into publishing *Origin of Species* by the parallel work of Wallace, he did so without mentioning human origins. On baring his soul to a close friend he claimed the revelation of his theory was "like confessing a murder."[5]

Nonetheless, exceptions were occurring. By Darwin's time in British intellectual life, moderate deistic and theistic opinion was gaining some hold if not full respectability. For example, some of David Hume's work was readily accepted. William Paley's influential *Evidences of the Christianity*,[6] while neither Trinitarian nor overtly reflecting Paley's deism, had become

2. Chadwick, *The Secularization of the European Mind in the Nineteenth Century*, 144–48.
3. Ibid.
4. Desmond and Moore, *Darwin*, 38.
5. Darwin, *Life and Letters*, 23.
6. Paley, *A View of the Evidences of Christianity*.

one of the standard theological texts set for the "little go" at Cambridge during and even after Darwin's student days.

The conditional acceptance of deism by the 1830's appears to have been dependent on factors such as: whether or not its expression was supportive of the established Church and the good ordering of society; the growing influence of recent scientific discoveries and theories particularly in Scottish and European geology; and the growing number of deist and theist sympathisers among the university trained and upper classes. Respectability seems to have involved avoiding potentially contentious theological notions. Divine agency in the world had become understood in a way that could be expressed as that of a generic perfect-being so that contentious matters of Christology and the Trinity could be avoided. Despite the intensity of expressed theological differences, all of the parties with a measure of acceptability in the early nineteenth century had at least two beliefs in common—an absolute trust in the inspired Scripture as the perfect work of God and the perfection of God's work being reflected in the created order.

As has been discussed, this second assumption provided a major impetus to natural philosophy, the domain of the amateur scientist-clergyman prior to the early nineteenth century. For this enthusiastic group, correctly reading the unsullied and perfectly designed book of nature in exacting detail became as much or even more important than reading the Bible. This was particularly the case for those influenced by Newton's arguments that the received text was suspect.

The notion of divine perfections in the world meant: there had to be cause for everything; God must always have a motive for acting and act for the best, and; the world is the best of possible worlds. It also followed in this logic that if God had to tinker with design this would indicate that the design was less than perfect. This could not be and so species could not be mutable. All things, all creatures, it was thought and widely accepted, had been perfectly created to suit their environments so that no divine adjustment was needed. These ideas found standard expression in Paley's *Evidences of Christianity* and *Natural Theology*.[7] The development of geology, palaeontology, biology and botany provide the academic context for understanding Darwin and Huxley. It is also necessary to describe Paley as he was influential in Darwin's early academic life. Paley's works emerge from the post Newtonian theological deist discourses and the natural theology movement. Darwin initially enthusiastically embraced Paley's description of Christian faith, but later explicitly disowned Paley and became agnostic about the faith.

7. Paley, *Natural Theology*.

Paley: Perfection Reflected in an Harmonious World

Paley had been influenced by Newton's ideas as presented by Samuel Clarke. Paley made regular allusions to Clarke's ideas[8] in his *Evidences of Christianity* which is a wide-ranging defence of the Christian faith. *Evidences* was written by Paley after Hume expressed scepticism arguing that miracles were inconsistent with perfect natural law.[9]

Originally published in 1794, Paley's three-volume work seeks to gain the readers assent to the Christian faith by appealing to its inherent reasonableness, to the evidence of the veracity of the scriptural miracles, and to the length to which their witnesses have gone to assert their truth. Paley points to the absolute dependability of revealed religion: "The gospels ... must be true as far as the fidelity of human recollection is usually to be depended upon, that is, must be true in substance, and in their principal parts, which is sufficient for the purpose of proving a supernatural agency."[10] Paley makes fidelity the test of Divine action. The Scriptures have their common character as the inspired word of God as a result of the witness of "holy men of God ... as moved by the Holy Ghost."[11] This human element of witness under duress and the effect of changed lives are for Paley the highest testament to the veracity of the scriptural accounts and thus to the divine origin of the texts.[12]

Natural Theology is the last of Paley's three works and provides a sustained argument for the nature and existence of God from an examination of the blessed harmony of created order. His argument begins by arguing that the existence of a watch implies a watchmaker; that the existence of Natural law implies a creator.[13] What Paley hints at in *Evidences*, he fully develops in *Natural Theology*. An essential conclusion in his argument is that the perfect God creating perfectly would not need to change anything, nor consequently would anything need to change in an essentially harmonious and happy creation. Therefore, species would logically be fixed and immutable. It is this Paleyan orthodoxy, which dominated both theological thought and the thought of the Royal Society's natural philosophers even during Darwin's early career.

---

8. Paley, *A View of the Evidences of Christianity*, I.XI, I.XIII.

9. Hume, "An Enquiry Concerning the Human Understanding," (1748), X.I.90; X.II.97.

10. Ibid. I.VIII.

11. Paley, *A View of the Evidences of Christianity*, Appendix B.

12. Ibid. XII Vol. 1 Proposition 2.2.

13. Paley's second book dealt with civil government.

By the 1830's, elements of this Paleyan orthodoxy were under challenge from a number of areas. One that does not affect the young Darwin were radical new theories developed in Tübingen under the influence of Baur and Strauss concerning the authorship of Genesis and the remainder of the Pentateuch. This development of what would evolve into higher criticism occurs too late for it to have any serious influence on Darwin's theological progress. The influence of this work develops in the English-speaking world about the time of Darwin's much delayed publication of *Origin* and later significantly informs Huxley's engagement with Christianity. As will later be discussed, Huxley with his usual polemic style blended a sceptic's version of higher critical scholarship into his denigration of established religion and promotion of evolutionary theory.

The second challenge to Paley came from the Empiricist philosophers. Hume's *Enquiry* directly countered Paley's *Natural Theology*. The highest form of proof for Hume becomes not that of Scripture attested to by reliable witnesses but rather that which is able to be tested by the senses.[14] Hume contended that the existence of a clock does not necessarily imply a watchmaker, arguing that the notion of prior cause depends firstly on human imagination. While Hume undermines Paley's argument for a first or ultimate cause, both he and Paley assume God is a perfect-being who acts perfectly. It is this linking of perfection of divine action to immutability that created problems with the growing body of data from geology and palaeontology. By the early nineteenth century, there was growing evidence for the extreme age of the earth, ice ages and extinctions which did not fit well within an assumed perfect changeless order. There was a growing realisation that there had been prehistoric changes in climate and that species may not in fact be fixed. This idea was contested even into the 1860's.[15] Even when variation in species was accepted, it was often assumed that there was still a divine plan. These nineteenth-century relationships between theology and science developed as a result of the affirmation, rejection or unquestioning acceptance of notions inherent in the non-christological understanding of divine agency in the world. A number of commentators have observed that the development of natural history and biblical geology contained the seeds of their own destruction and inherent self-contradiction. In the case of the notion that the world reflected the perfection of a divine perfect-being helped to lead to the development of the very exacting methods of detailed study which found data that contradicted this assumption. The extent of the flaw in this assumption is highlighted when the Paleyan harmonious

---

14. Hume, "An Enquiry Concerning the Human Understanding," V.II.45.
15. Desmond and Moore, *Darwin*, 478–90.

blessed natural order is rejected by Darwin as he developed his theory. Darwin becomes an example of a scientist who found himself in a quandary regarding the Christian faith as he proved its "precondition" false. Therefore as a precondition to the Christian faith, perfection of divine agency became a stumbling block to interaction between theology and science.

Darwin: Perfection No More

The change in Darwin's thinking from his initial Paleyan orthodoxy to his later agnosticism or theism is not a simple process. Darwin was influenced in the development of his own thinking and research and by his teachers at Edinburgh and Cambridge. His eventual rejection of immutability as an expression of the divine perfections in nature in Paleyan guise led him into an intellectual opposition to what were then assumed to be the preconditions to the validity of the Christian faith. His loss of faith however was occasioned by his concerns over his father's death and finally the death of his beloved daughter Annie. While personal tragedy led to his giving up Christianity, he rationalized this step. Ospovat has conducted a detailed study of Darwin's notebooks tracing many influences on his developing theory, theological, scientific and social. The development of Darwin's thought is complex but observable in his extant copious notes and revisions of essay material. The current chapter deals with Paley's influence on Darwin's thought.[16] Darwin's explanation shows his investment in the Paleyan description of Christianity, which assumed that divine existence and action and providence is dependent on the perfection of the agency of an omnipotent God. Darwin's intellectual justification for his prior rejection of traditional Christianity is that his research showed this precondition to be false.

Darwin's religious views never evolved into a definitive position and he was usually reluctant to speak or write of it. In his own notes he identifies the variations and at times the vagueness of his theological thinking. As will be shown, Darwin was aware of the nature of the theological arguments, but believed he was limited to choosing between Paley's rational biblical perfection on the one hand and Coleridgean experientialism on the other. Recent commentators such as Brooke, Barbour, and Harrison demonstrate the complexity of the relationship between science and religion in Victorian England, including varying levels of support for different theories. One area of conflict is without doubt the conflict between Darwin's thinking and what in the early nineteenth century were conventionally held to be

---

16. Ospovat, *The Development of Darwin's Theory*, 60–86.

preconditions to the Christian faith. Darwin lived with this internal conflict most of his adult life.

Darwin the Paleyan

Darwin wrote that when his father "proposed that I should become a clergyman. I did not then in the least doubt the strict and literal truth of every word in the Bible; I soon persuaded myself that our Creed must be fully accepted."[17] He read the conservative theologian Sumner's *Evidences of Christianity* in preparation for the possibility of undertaking Anglican orders at Cambridge. Finding nothing objectionable in its arguments, he was able with integrity to enrol at Cambridge.[18] In good conscience he was able to sign the 39 articles prior to his graduation.[19] While at that time he may have agreed with a conventional Christianity, he felt no inward religious conviction of call to holy orders.[20] He was content even if his religious sentiments were weak, that his confidence in revelation was assured by God's perfect action in creation. He depended on the strength of Paley's argument. In his first year at Cambridge 1830 staying in Paley's old rooms, he absorbed Paley and was convinced that he could, if needed, have written out Paley's *Evidences* from memory.[21]

He remained a committed Paleyan out of university and onto the *Beagle*. He commented on the Paleyan harmonious perfection of nature shown in the happiness of man and beast at the zoological gardens.[22] Darwin trusted Scripture and the argument for God's existence from the blessed and harmonious creation. Darwin himself states that he had no doubts about the "Unanswerable authority" of the Scripture during his time on the *Beagle*.[23] Nonetheless, Paley's particular arguments on immutable fixity of species had not convinced Darwin, who alluded to the creation of species while investigating beds of shells found 400m above sea level in South America.[24] He was later more explicit about mutability in his notes commenting that Lyell was right about the "gradual birth and death of species."[25] But he re-

17. Darwin, *Autobiography*.
18. Desmond and Moore, *Darwin*, 48–49.
19. Ibid., 93.
20. Desmond and Moore, *Darwin*, 64, 66.
21. Darwin, *Autobiography*.
22. Darwin, *Correspondence*, 1:121–22.
23. F. Darwin, *The Autobiography of Charles Darwin and Selected Letters*, 62.
24. C. Darwin, *Diary* 23/7/1834, in Desmond and Moore, *Darwin*, 155.
25. Darwin, "The position of the bones of Mastodon (?) at Port St Julian is of

stated his commitment to the Paleyan notion of right order expressing the view that, successive births must repeople the globe to keep the harmony established by the "Author of Nature."[26] Indeed, he argued a Paleyan case on common design for ant lions in different continents. That is—that one perfect design implied one perfect designer.[27] Upon his return to England, the diligent Darwin found himself gaining respect among the leaders of science of his time. However, the more he reflected on his data from the *Beagle*, the more he became committed to explaining how it was that species developed, and in the process he came to a point at which he abandoned Paley. Thus, he found himself in a strange situation developing a theory to explain the mutability of species, whilst becoming a leading member of the established scientific fraternity who despised such notions.

Doubts

It is well established that Darwin had doubts about the nature of faith during the late 1830's as he began to think about species mutability. Unlike Newton's works which are often undated or difficult to date, Darwin's correspondence and diaries enable the researcher to trace the development of his thought over time. Darwin was very cautious about who he shared his thoughts during this period. His cousin Emma, later his wife, and friend Hensleigh were among the first outside his immediate family to have his confidence. He was a committed family man who dearly loved his children. Ironically, as he discovered the laws of inheritance, he worried about their health as offspring of a first cousin marriage to his wife Emma.

He began to reassess the theology of perfection in his notes on descent and here he began to see that perfection in nature might be serendipitous, that fitness might be the result of chance.[28] This is not yet survival of the fittest, for this comes only after he begins to apply Malthusian ideas to his own work and begins to grasp the darker side of nature. Ospovat indicates that the change was sudden after Darwin read Malthus, though the change in his notebooks "did not occur until some weeks after."[29] The content of three letters sent to Darwin by Emma during their courtship and early in

---

interest" (CUL-DAR42.97–99, Feb 5 1835), 2r

26. Ibid., 2v.

27. Darwin, *Correspondence*, 1:481.

28. Darwin, *Notebook C61*, in P. H. Barrett, "Darwin's Early and Unpublished Notebooks," 448.

29. Ospovat, *The Development of Darwin's Theory*, 69; Darwin, *Correspondence*, 6:178.

their marriage is instructive in relation to what they indicate about Charles' assumptions. Darwin's doubts about Paleyan theology of his Cambridge days began to be expressed in his notes during 1838, the same period as these letters.

That both Charles and Emma held the assumption that the alternative to basing faith on rational evidence was to be found in the religious sentiments, is evidenced in Emma's appeal to the emotional impact of a particular biblical passage and her confidence that as newlyweds they shared the same sentiments about personal religion.

> My reason tells me that honest & conscientious doubts cannot be a sin, but I feel it would be a painful void between us. I thank you from my heart for your openness with me & I should dread the feeling that you were concealing your opinions from the fear of giving me pain. It is perhaps foolish of me to say this much but my own dear Charley we now do belong to each other & I cannot help being open with you. Will you do me a favour? Yes I am sure you will, it is to read our Saviour's farewell discourse to his disciples which begins at the end of the 13th Chap of John. It is so full of love to them & devotion & every beautiful feeling. It is the part of the New Testament I love best.[30]
>
> Charles there is only one subject in the world that ever gives me a moment's uneasiness & I believe I think about that very little when I am with you & I do hope that though our opinions may not agree upon all points of religion we may sympathize a good deal in our *feelings* on the subject. I believe my chief danger will be that I shall lead so happy comfortable & amusing a life that I shall be careless & good for nothing & think of nothing serious in this world or the next. However I won't be solemn either.[31]

Emma's was not an anti-rationalist appeal to religious sentiment as shown by her continuing confidence in the outcome of honest and conscientious exploration of doubts, given the limits of human understanding.[32]

Darwin was considering rejecting the notion of revelation. Darwin had accepted Paley's explanation of the Christian faith that revelation in Scripture was attested to by the perfection of God's revelation in nature. Darwin was coming to believe the later did not hold. If the Unitarian Lyell or the more conservative Sedgwick, Owen and Whewell had known Darwin's private thoughts on transmutation, they would not have given him the

30. Ibid., 2:122.
31. Ibid., 2:169.
32. Ibid., 2:171.

cordial reception and support that they did, having publicly attacked such notions.[33] Nonetheless, Darwin sat in the secretary's chair of the Geological Society as his old teacher Grant had his Lamarckian views dismembered in 1838. Darwin was able to salvage some ideas from the debacle for transmutation. He no longer accepted that there could be an absolute tendency to progression as Lamarck had hoped[34] or for notions of planned design. There had to be some other mechanism. Darwin's hypothesis of change by natural selection in the struggle for survival was only one of a number of competing theories offered during the nineteenth century to explain change in species and extinctions. The natural selection model for evolution did not enjoy immediate and widespread assent even in the scientific community. The debate over the mechanism for evolution has continued well into the Twentieth century.[35]

In August of 1838 Darwin became taken with the substance of Comte's *Positive Philosophy*.[36] Darwin was particularly impressed by Comte's description that mature science was characterized by belief in the rule of law. Darwin left active agency of divine providence in the process behind in favor of discovering some constant law. An atheist, Comte was interested in tracing facts to laws and was not interested in divine action. Darwin saw in Comte's scheme a grander view of nature than the notion of God individually crafting each lowly slug and snail. He writes, "How beneath the dignity of him!"[37] Later in 1842 in drafting a theory of natural selection, this arbitrary idea of perfection emerges: "Wild animals are not a product of God's whim any more than planets are held up by his will. Everything results from grand laws—laws that 'should exalt our notion of the power of the omniscient Creator.'"[38] What were those laws? There arises a similar question regarding how to define dignity arises as it does for defining perfection. What is below God's dignity? Ultimately this question becomes just as problematic as asking whether God would be involved with anything less than perfect? This view is not consistent with incarnation. What for example could be less dignified or less than perfectly powerful than God as a baby in a nappy in the nativity.

33. Desmond and Moore, *Darwin*, 239, 413. Darwin, *Correspondence*, 1:540, 3:68.

34. Desmond and Moore, *Darwin*, 274-75.

35. An excellent outline of the progress of debate surrounding and the varying fortunes of Darwinism is Bowler, *Reconciling Science and Religion: The Debate in Early Twentieth Century Britain*.

36. Comte, *The Positive Philosophy of Auguste Comte*.

37. Darwin, *Notebook D37*, in P. H. Barrett, "Darwin's Early and Unpublished Notebooks," 455.

38. Darwin, *Notebook MAC*, in Desmond and Moore, *Darwin*, 293.

Darwin, even in late 1844, tied a notion of perfection to divine action in law and purpose. As Ospovat states, "As Late as 1844, the structure of Darwin's theory was to a large extent determined by . . . natural theological ideas or assumptions that Darwin had held since before opening his first transmutation notebook."[39] Ospovat traces Darwin's movement from holding transmutation as a progression of perfect forms to transmutation through relative change.[40] This change encompasses a profound theological shift for Darwin from that of transmutation occurring as planned with divine purpose to transmutation occurring by chance. This change occurred after Darwin assimilated Malthus. In late September of 1838, Darwin read the sixth edition of Malthus' *Essay on the Principle of Population* with its description of the weak succumbing in the struggle for available resources.[41] Darwin describes this as very timely influence. As Ospovat indicates, Paley had managed to encapsulate population pressure in his harmonious system.[42] That is that population pressure led to checks and balances which enhanced harmonious perfection. Counter to Paley, Darwin came to realize that disharmony is "the necessary concomitant of transmutation by means of the warring species."[43] Reflecting on the concept of inheritance in the light of Malthus, Darwin came to scorn the common idea that God had created rudimentary parts like male nipples, the human coccyx or tail. "What bosch!! . . . The designs of an omnipotent creator exhausted . . . such is Man's philosophy when he argues about his creator!"[44] This argument was to be later polished and incorporated in *Origin*.[45] Paley was now left far behind.

Darwin realized that he was going far beyond Lamarck, whom Sedgwick and Whewell despised, and consequently he was deeply concerned about the hysteria his views would unleash.[46] He began to broach his theories with close confidants such as Hensleigh. In a letter to Hensleigh, his close colleagues Lyell and Hooker were shaken when informed about his work on species, which Darwin took as a further caution.[47] Nevertheless, Darwin constantly sought their advice and tested his ideas on them. Huxley

---

39. Ospovat, *The Development of Darwin's Theory*, 2.
40. Ibid., 61–64.
41. Darwin, *Notebook MAC*, in Desmond and Moore, *Darwin*, 264.
42. Ospovat, *The Development of Darwin's Theory*, 64–66.
43. Ibid., 67.
44. Darwin, *Notebook Summer 1842*, in Desmond and Moore, *Darwin*, 272.
45. Darwin, *Origin*, XIII; Young, *Darwin's Metaphor: Nature's Place in Victorian Culture*, 23–55.
46. Desmond and Moore, *Darwin*, 249.
47. Darwin, *Correspondence*, 5:155, 294, 379; Darwin, *Correspondence*, 6:516–17.

was later added to the inner group of confidants during the preparation of *Origin*.[48]

At the end of his life Darwin stated that Paley's argument from design which had at first seemed so convincing, failed in the light of the discovery of the law of natural selection.[49] Whatever the serious questions Darwin had about traditional Christianity as he had learnt it, this did not precipitate his loss of faith. That followed personal tragedy.

Departing the Faith

In 1850, dreading his father's imminent death, Charles and Emma explored the religious journeys of others including the influential Coleridge and the dry logic of the "Unitarian Pope" Norton.[50] Emma longed for Charles to find hope in the promise of eternal life. Nevertheless, Darwin was unmoved by Coleridge's appeal to innate religious feeling. Darwin was too grounded in the idea that the preconditions for Christianity were to be found in the evidence of perfect divine agency in the world as argued in the works of Paley and Norton. Darwin still believed in a creator and he still thought highly of his Coleridgean colleagues such as Owen, but did not share their idealism.[51] Later Darwin would write,

> At the present day the most usual argument for the existence of an intelligent God is drawn from inward conviction and feelings which are experienced by most persons. Formerly I was led by feeling . . . although I do not think that the religious sentiment was ever strong in me. Now the grandest scene would not cause any such convictions and feelings to rise in my soul.[52]

Darwin found no solace in Coleridge as his father died. He remained concerned about his father's eternal fate and the "damnable doctrine of perdition."[53] His daughter Annie's death in 1850 was the final point of departure from his faith. Francis Darwin in his *Reminiscences* repeats the words

---

48. Darwin, *Autobiography*, 227.

49. Ibid., 63.

50. Coleridge, *The Friend and Aids to Reflection*; Norton, *The Genuineness of the Gospels*; Desmond and Moore, *Darwin*, 358–60.

51. Darwin, *Correspondence*, 4:127, 219.

52. Darwin, *Autobiography*, 65.

53. Brooke and Cantor, *Reconstructing Nature: The Engagement of Science and Religion*, 31; Desmond and Moore, *Darwin*, 379–87.

his father wrote soon after. "We have lost the joy of the household."[54] Annie was not the first child he and Emma lost but she was the closest to him. The poignancy and tenderness and love of this man for this special daughter who was the delight of his heart cannot be overstated. Darwin, already in chronically poor health, slipped into an extended period of depression following the loss of these close family members. Annie's death put an end to Darwin's belief in a moral and just universe. Later he would say that this period chimed the final death-knell for his Christianity, even if it had been a long, drawn out process of decay. In publishing Darwin's views on religion, Aveling quoted him as saying "I never gave up Christianity until I was forty years of age." Charles' son Francis believed Aveling's report was accurate.[55] Though Charles Darwin would never describe himself as an atheist, in his autobiography he claims that "agnostic" would perhaps be the best description of his views, and then later "theist."[56]

Publicly Departing Paley

In between bouts of poor health Darwin continued his massive work on barnacles, largely leaving natural selection and theological speculations to one side.[57] He did not return to evolution until receiving Wallace's twenty page letter in 1848 which outlined a similar theory to Darwin's species change by natural selection, which led to Darwin writing and publishing *Origin* and publicly parting with Paleyan perfection and teleology.[58] Darwin's theory of evolution depended on natural selection as the species making mechanism. It is the removal of the necessity of divine or any other external uplifting force which distinguishes Darwin's theory from other "Paley and Co" evolutionary theories mooted at the time.[59] The potentially secular nature of the explanation and the extent of its departure from Paley's blessed harmony in nature were not lost on Darwin. "Nature's depravity cried out against a noble Providence. What a book a Devil's Chaplain might write on the clumsy, wasteful, blundering low and horridly cruel works of nature!"[60]

---

54. Darwin, *Autobiography*, 88–89.

55. Darwin, *Correspondence*, 5:32, 540–42; Desmond and Moore, *Darwin*, 387, 658; Darwin, *Autobiography*, 69.

56. Ibid., 66.

57. Darwin, *A Monograph on the Sub-Class Cirripedia*.

58. Desmond and Moore, *Darwin*, 468–70.

59. Darwin, *Correspondence*, 8:258.

60. Ibid., 6:178.

Ironically, the multiplicity of Galapagos finches and giant tortoises often cited as classic examples of natural selection were entirely missed by Darwin as separate species when he was there with the *Beagle*. As a result he failed to collect sufficient samples. In order to find sufficient evidence for his natural selection, he had to work with data he collected at home. Thus, Darwin became the first scientist to investigate domestic animals seriously.[61] *Origin* contains copious examples involving pigeons and various domestic animals as the key evidence for Darwin's theory. He became a frequenter of county shows and pigeon clubs. The ability to vary species under domestication as in horses, he claimed makes a, "mockery of the notion of the Works of God, aligning a view of the fixity of species with those 'unlearned' people who believed that fossils have never lived."[62]

However, in *Origin* he is still sympathetic to Paley while turning him on his head, "No organ will be formed, as Paley has remarked, for the purpose of causing pain or for doing an injury to its possessor."[63] This conciliatory phrasing is characteristic of *Origin*, which was no more unorthodox than the nature of the book could make it. Indeed, Darwin explicitly avoided any discussion of human origins or descent.[64] In summarising his proposed law of variation in *Origin*, Darwin still uses Paleyan terminology of beautiful and harmonious nature.

> Although new and important modifications may not arise from reversion and analogous variation, such modifications will add to the beautiful and harmonious diversity of nature. Whatever the cause may be of each slight difference between the offspring and their parents—and a cause for each must exist—we have reason to believe that it is the steady accumulation of beneficial differences which has given rise to all the more important modifications of structure in relation to the habits of each species.[65]

Darwin did not expect anybody to read *Origin*, and was genuinely surprised when the original print run of *Origin* was oversubscribed and was widely praised.[66] There was as he expected criticism from close friends. Owen argued for the ordained birth of species against Darwin's design by natural selection and chance. Sedgwick claimed variation was at the expressed causation of God. He also commented that Darwin's phrasing

61. Desmond and Moore, *Darwin*, 426.
62. Darwin, *Origin of Species*, V.
63. Ibid., VI.
64. Darwin, *Autobiography*, 209.
65. Darwin, *Origin of Species*, 195.
66. Darwin, *Autobiography*, 191.

of natural selection in *Origin* is as though it was almost occurring by a conscious agency.[67] Darwin's theory was greeted with Unitarian approval and even enthusiasm among Huxley's generation. However, the many in older generation of patrician Anglicans still feared that a nature not actively upheld by God's word, boded ill.[68] Although Darwin had departed Paley, not all his supporters thought this. Geologist and congregational deacon, Asa Gray in the United States wrote a supportive review of *Origin*.[69] Gray explicitly links Darwin and Paley, seeing no disparity between Paley's description of God's purposes in creation and Darwin's natural explanation of this by natural law.[70]

Darwin was impressed enough to provide half Gray's publication cost.[71] "Impressed" however should not be confused with "agreed with." In 1860, in letters to Gray, Darwin cited freak accidents and parasites in nature as inconsistent with the actions of a God of providence and rejected providential design in nature.[72] This is a recurring theme in his letters. Writing to one M. E. Boole a few years later in 1866, Darwin humbly claimed a layman's opinion:

> [I]t has always appeared to me more satisfactory to look at the immense amount of pain & suffering in this world, as the inevitable result of the natural sequence of events, i.e. general laws, rather than from the direct intervention of God though I am aware this is not logical with reference to an omniscient Deity.[73]

Darwin's struggle indicates that he subscribed without question to the notion that divine agency in the world if real must be perfect. Nature was not perfect: consequently God did not seem to act, leaving him with his "unresolvable conclusion." Writing to Sir John Herschel, Darwin expressed confusion about God acting in a fashion described by Paley and Leibniz. Not only was God expected to act perfectly God was not expected to act without some purpose for that action becoming apparent.[74]

Ultimately, variation did not appear to always be for the best for either the creature or nature in general. If divine perfection was not reflected in

---

67. Desmond and Moore, *Darwin*, 478–79; Darwin, *Autobiography*, 229.
68. Desmond and Moore, *Darwin*, 241, 233, 487–88.
69. Gray, *Natural Selection Not Inconsistent with Natural Theology*.
70. Gray, "The Origin of Species by Means of Natural Selection."
71. Desmond and Moore, *Darwin*, 502.
72. Darwin, *Autobiography*, 67, 247, 249.
73. Darwin, "Letter 5307—Darwin, C. R. to Boole, M. E., 14 Dec 1866."
74. Darwin, *Correspondence*, 9:135.

perfect variation was perfection reflected in nature's purpose? Darwin had assumed God's perfect purposeful action must be extended into perfect prescience for there to be perfect law enabling teleological purpose and enduring providence. This too was problematic for Darwin for the same reason—nature appears to just happen without a reason.[75]

In spite of this negative assessment, Darwin, in what is for him a rare public foray into theological speculation, actively engages with Gray in the conclusion to Darwin's seminal work *The Variation of Plants and Animals under Domestication*.[76] Darwin's conclusion to *Variation under Domestication* begins with an acknowledgement of the criticism that he has not proposed a mechanism for species change.

> Some authors have declared that natural selection explains nothing, unless the precise cause of each slight individual difference be made clear. If it were explained to a savage utterly ignorant of the art of building, how the edifice had been raised stone upon stone, and why wedge-formed fragments were used for the arches, flat stones for the roof . . . But this is a nearly parallel case with the objection that selection explains nothing, because we know not the cause of each

The answer Darwin provides is that although the scientific investigator only has access to the random fragments or end result of change, what seems random or even arbitrary is the result of natural law.[77] Cautiously acknowledging that he is moving to questions of theology rather than science, Darwin questions why should a God impose a natural law in such a fashion that happens to fit some arbitrary human aesthetic?

> And here we are led to face a great difficulty, in alluding to which I am aware that I am travelling beyond my proper province. An omniscient Creator must have foreseen every consequence which results from the laws imposed by Him. But can it be reasonably maintained that the Creator intentionally ordered, if we use the words in any ordinary sense, that certain fragments of rock should assume certain shapes so that the builder might erect his edifice?[78]

---

75. Ibid., 9:225.
76. Darwin, "The Variation of Plants and Animals under Domestication," 427–29.
77. Ibid., 428.
78. Ibid.

Furthermore, Darwin asks why God should have acted with this foresight to allow purposeless variation or even the possibly of harmful variant.[79] How could God allow that humans may use variation to produce deformation, ugliness, evil and pain in contrast to healthy perfect species? "Did He ordain that the crop and tail-feathers of the pigeon should vary in order that the fancier might make his grotesque pouter and fantail breeds?"[80]

If there is some divinely guided natural law of selection it must also perfectly provide for these negative variations as well as for those of all species "most perfectly adapted, man included."[81] The combination of the issues of apparently absent omniscient prescience, apparently purposeless variation and copious opportunities for pain and suffering, contribute to Darwin's agnosticism. He could not see how to affirm in any way the agency of a divine perfect-being in the world. Therefore, while Darwin held doubts about Gray's adaptation of Paley, these doubts leant toward scepticism.

> However, much we may wish it, we can hardly follow Professor Asa Gray in his belief "that variation has been led along certain beneficial lines," like a stream "along definite and useful lines of irrigation." If we assume that each particular variation was from the beginning of all time preordained, then that plasticity of organisation, which leads to many injurious deviations of structure, as well as the redundant power of reproduction which inevitably leads to a struggle for existence, and, as a consequence, to the natural selection or survival of the fittest, must appear to us superfluous laws of nature. On the other hand, an omnipotent and omniscient Creator ordains everything and foresees everything. Thus we are brought face to face with a difficulty as insoluble as is that of free will and predestination.[82]

Darwin struggled with what he saw as the valid concerns of Huxley and Lyell and continued to be worried that he might actually be wrong.[83] Darwin continued to be worried because he could not determine a mechanism for change even though he was sure variations occurred. This was something that Huxley often quizzed him on, and remains the reason why evolution remains technically a scientific theory rather than an established law even to the present. It is also why Darwinian evolution has suffered varying fortunes

---

79. Ibid.
80. Ibid.
81. Ibid.
82. Ibid., 429.
83. Desmond and Moore, *Darwin*, 510; Darwin, *Correspondence*, 8:260

during the twentieth century.[84] While his doubts later diminished, Darwin was never able to fully eliminate them to his own satisfaction.

Darwin saw his proposed law of natural selection as constant and universal. It was his caution in relation to public sensibility which led him not to draw conclusions regarding human development in *Origin*. It was only a matter of time before the theory of natural selection was extended to humans. Wallace's first paper on this topic was given in 1864 to the ultra-racist pro-slavery Anthropological Society. In contrast to Wallace's audience Huxley, the future X club and Darwin were abolitionist and opposed slavery.[85] Darwin finally published *Descent of Man* in 1871, tying natural selection to human evolution. This publicly broke his ties between the notion of perfect divine providential design in human development and Coleridge's notion of innate religious feeling or belief. "There is no evidence that man was originally endowed with the ennobling belief in the existence of an Omnipotent God."[86] *Descent* does not however depart from belief in God but rather made evolution of religion the highest form of human development. "The idea of a universal and beneficent Creator does not seem to arise in the mind of man, until he has been elevated by long-continued culture."[87] His thesis however overturned traditional proofs of God's existence.

> I am aware that the assumed instinctive belief in God has been used by many persons as an argument for His existence. But this is a rash argument, as we should thus be compelled to believe in the existence of many cruel and malignant spirits, only a little more powerful than man; for the belief in them is far more general than in a beneficent Deity.[88]

Darwin builds on his earlier discussion with Hensleigh, arguing if there were to be divine revelation then it would have to be in common for all people.[89] Darwin still saw religious belief as important, but that it arose from evolutionary development along with morality.[90] He tied fear of reprobation and morals to a natural selection argument rather than as arising from divine agency in the world.[91]

84. Bowler, *The Eclipse of Darwinism*.
85. Desmond and Moore, *Darwin*, 521.
86. Darwin, *The Descent of Man*, III, XI.
87. Ibid., XI.
88. Ibid., XI.
89. Ibid., IV.
90. Ibid.
91. Ibid., IV, V.

Although Darwin used theological language freely in *Origin* and discussed the evolution of religion in the *Descent of Man*, he was very discreet about revealing his personal beliefs. As his son reported, he admitted to his family, "Science has nothing to do with Christ, except in so far as the habit of scientific research makes a man cautious in admitting evidence. For myself, I do not believe that there ever has been any revelation."[92] Later in life he regretted using this type of language in *Origin*, as he wrote to Hooker: "But I have long regretted that I truckled to public opinion & used Pentateuchal term of creation, by which I really meant 'appeared' by some wholly unknown process."[93] He also was much blunter on occasion in his private correspondence than his son Francis intimated. Responding to a private inquiry about his religious opinions Charles Darwin replied "I am sorry to have to inform you that I do not believe in the Bible as a divine revelation, & therefore not in Jesus Christ as the son of God."[94] Darwin usually hesitated, however, to speak publicly about his theological views. Darwin viewed Huxley as more eloquent and better able to argue what he saw as the cause of truth.[95] Nevertheless, when drawn by trusted colleagues, he did express his views in print. One such example was a letter sent by Darwin to Asa Grey and B B Warfield. Warfield noted Darwin's affinity with Huxley's religious views and Darwin's own affirmation of agnosticism. In publishing the letter, Warfield also referred to local journal article by Darwin.[96] In this letter, Darwin is explicit about not warming to religious feeling. He indicates that he had neither the time nor the energy to explain this fully; this is not surprising given the state of his health in this later period of his life. What stands out however is that Darwin voiced almost the same ideas as Huxley in deriding faith in the spiritual nature of humans, even referring people to Huxley's work on nerves and the brain to ask the agnostic question, "Does the human spirit exist and if so where?"[97] Darwin saw that the best weak argument for God then lay in "the religious sentiment." Then again, like Huxley, his research led him to think these feelings and intuitions may have merely resulted from the chance development of biological processes rather

---

92. Darwin, *Autobiography*, 61

93. Darwin, *Correspondence*, 11:277.

94. Darwin, "Darwin, C. R. To McDermott, F. A., 24 Nov 1880."

95. Darwin, *Autobiography*, II.

96. Darwin, "Darwin to Grey," 11/3/1878, in Noll and Livingstone, *B B Warfield: Evolution, Science and Scripture, Selected Writings*, 111.

97. Reference will be made later to Huxley, "Has a Frog a Soul?"

than from any divine agency. Hence he too, like Huxley, found he must "leave the problem insoluble."[98]

Darwin's rejection of traditional Christianity was grounded in personal tragedy. He did however, rationalize this rejection in the incompatibility of his research with what he had assumed faith must depend on. His rationalisation is explicitly a rejection of the Paleyan notion of perfect harmonious design as proof of the Christian faith. Darwin does not examine the underlying assumptions himself. In the end, Darwin, while a trained cleric, is not so much a theologian but one of the first of the new professional scientists, a generation early. All his life, his primary interest remained the description of nature. He was more than content to leave debate to those he considered more capable at handling the intensity of open controversy—such as Huxley. Darwin actively supported Huxley and largely agreed with his ideas on religion, contributing some ideas to Huxley in correspondence. It is Huxley however, who explicitly ties his rejection of the Christian faith and divine agency in the world in part to the Augustinian description of inspiration and its dependence of the existence of the soul.

Huxley: Metaphysics No More

Huxley, unlike the retiring Darwin, was an active and not unbiased promoter of Darwin's ideas as well as his own. Whereas Darwin tended to avoid controversy, Huxley felt he could respond well with quick repartee. Huxley revelled in the cut and thrust of debate, Darwin believed he had the dash and verve which knew he lacked. Darwin actively looked to Huxley as an energetic promoter of the truth as he saw it.

Huxley's reasons for reacting against the established Church and traditional Christianity are interesting. "The man who believes on human testimony that a virgin bore a child, and that a dead man came back to life, is a superstitious creature who would believe anything."[99] Particularly as he notes the extent of his earlier biblical literalism, "I, too, began life . . . with implicit faith in the Bible, or, rather, in what I was told the Bible taught. I, too, supposed that the world was created in 144 hours 6,000 years ago."[100] He became one of the anti-establishment radicals who, during Darwin's younger days, had been the source of concern to the ruling classes. However, a generation later, Huxley's radicalism merely landed him in controversy

---

98. Darwin, "Darwin to Grey," 11/3/1878, in Noll and Livingstone, *B B Warfield: Evolution, Science and Scripture, Selected Writings*, 111.

99. Huxley, "The Bible and Modern Criticism (in the Times)," 4.

100. Ibid.

rather than in jail. Also, having been raised in a dissenting lower class family he also had to deal directly with Anglican and academic prejudice.

He faced difficulty with university level education as the abolition of religious tests for Oxford and Cambridge did not occur until 1871.[101] He had little tolerance for shallow thinking and prejudice and remained sharply outspoken and controversial.[102] Together with other close colleagues, he was intent on stamping out the Royal Society's old amateur ethos and helped to usher in the era of the professional scientist. Those of his set did this by gaining teaching positions as well as positions of influence. Huxley was appointed to the Royal School of Mines in 1854 and he used that position and the influence of his close colleagues to successfully keep debate in the public arena. Together with other close colleagues, he was intent on stamping out the Royal Society's old amateur ethos and helped to usher in the era of the professional scientist. He maintained a lively and academic interest in biblical and theological studies and engaging theologians on a number of occasions with most recent scholarship. Huxley explicitly tied the doctrine of inspiration to his reasons for rejecting traditional Christianity.

Antiestablishment

His opposition to traditional society and religion was not "radical" as in overturning the order of society, but rather in reforming it. He still valued faith, though not the Church which he saw as corrupt and obscurantist. Nevertheless he married, had his children baptized (Unitarian), promoted bible reading and public education. His motivation was always the pursuit and the promotion of truth—truth that is demonstrable by scientific reason. He was much taken with the sentiment of positivism and referred warmly to the work of Comte in his writing and in his argument in "An Apologetic Irenicon."[103] Adding positivism to his toolkit, he was determined to end the amateur status of science in the academy, considering that there was no place for anything less than solid clear thinking based on truth not censored by prior commitment to Church or societal order. Science deserved the unbridled truth and nothing less. There are, he argued,

> indeed, some who seem to suppose that the infallible Church guarantees the infallibility of the Bible; and that the infallible Bible guarantees the infallibility of the Church. But, if the famous

---

101. Latourette, *The Nineteenth Century in Europe*, 2:315.
102. Desmond and Moore, *Darwin*, 411, 433.
103. Huxley, "An Apologetic Irenicon in the Fortnightly Review," 5:557.

Hindu who rested the earth-bearing elephant upon a tortoise and was met by the question, "On what then does the tortoise rest?" had answered, "On the elephant," the reply would not have very much assisted the querier. And I think the argument that since X says Y is infallible and Y says X is infallible, therefore both are infallible, will as little satisfy any one accustomed to the use of a logic whose tracks are not circular.[104]

It is logic and reason that shape Huxley's hostility to the establishment. Why should authority continue to flourish merely based on tradition, particularly if that tradition were illogical? Huxley argued that the tradition of free thinking to which he claimed to belong had a long history and that only in free criticism of all things—including theology—could reason and the truth be set free from artificial limits. That science and religion must agree, he saw as both unnecessary and complicating the open search for truth. He argued that "Theology and Parsondom are the irreconcilable enemies of Science," and saw transmutation of species as a tool to drive a wedge between theology and science.[105] Huxley rejected the notion of the two books and considered divine perfections irrelevant to the study of nature. As already noted, transmutation was strongly opposed by the scientific aristocratic establishment who in the mid-nineteenth century led and largely controlled science. Huxley sought and found ready allies to promote the cause of the truth outside the establishment of the Royal Society. the "X" club.[106]

Gorillas, Descent, and Lay Sermons

Many of the pictures that still exist of Huxley in his earlier years typically show him sitting or lecturing with a gorilla skull. Whereas Paley had assumed that similarity in anatomical design was evidence of God's use of one perfect design, Huxley argued that this same similarity was the evidence of common ancestry of species. He argued strongly that his hearers should put aside their preconceptions, because similarity of structure in ape and human skeletons made the point obvious.

There were, however, contrasts between Huxley's actions and methods and those of Darwin. Some were aghast that at the same time Huxley's lay sermons were turning Congregationalists into Pantheists, Huxley was

---

104. Huxley, "The Bible and Modern Criticism (in the Times)," 15.

105. Desmond and Moore, *Darwin*, 465, 472; Desmond, *Huxley: The Devil's Disciple*, 253.

106. Barton, "An Influential Set of Chaps," 53–81.

campaigning to teach children Scripture.[107] While unexpected, Huxley had a particular interest in Scripture as attested to by the number and the erudition of the essays he wrote on scriptural topics. While eschewing traditional belief, Huxley, like Darwin, viewed religion as one of the high points of human development, though in need of further reform on a scientific basis. Huxley drew a contrast between what he believed was true and what he valued in his "Lectures on Evolution." Huxley claimed that the long-term duration of world history had an extensive pedigree. He maintained that such a view "was held more or less distinctly, sometimes combined with the notion of recurrent cycles of change, in ancient times; and its influence has been felt down to the present day. . . . Uniformitarianism, with which geologists are familiar. That doctrine was held by Hutton and in his earlier days by Lyell."[108] Nevertheless, Huxley also describes positively the influence of the doctrine of sudden origin,

> which you will find stated most fully and clearly in the immortal poem of John Milton–the English Divina Commedia– "Paradise Lost." I believe it is largely to the influence of that remarkable work, combined with the daily teachings to which we have all listened in our childhood, that this hypothesis owes its general wide diffusion as one of the current beliefs of English-speaking people.[109]

While advocating that history was epochs long, Huxley gave honor to Milton highlighting the important place he considered religious thought had in civilized society and even in his own upbringing. Huxley's style—though usually more polemic than this last example—fits the usual pattern of popularisers of his time. Indeed, Huxley was so popular that his Saturday night public lectures, the lay sermons, as they became known, were delivered to packed houses and gained him continuing notoriety among the set of Owen and Sedgwick. This popularity arose even before Huxley came to accept natural selection as a mechanism for evolution; Huxley had already rejected assumptions which were part of the old conventional way of describing the world and the nature of God's action in it.[110]

---

107. Desmond, *Huxley: Evolution's High Priest*, 21.
108. Huxley, "Lectures on Evolution," 4:51.
109. Ibid., 52.
110. Gillespie, *Genesis and Geology*, 24.

Promoting Darwinism

Natural law in some form, Huxley was sure, applied to evolution. It would be a principle which, universally applied, would explain the development and variation of all species. Huxley, while unsure of the mechanism, was certain that evolution was a historical fact. So he promoted evolution initially without natural selection, but also without reference to natural history or theology.[111] When he read *Origin* prior to its publication, he reacted positively to it, though he had to read it many times to let the volume of information sink in and to grasp adequately the theory Darwin was promoting. *Origin* is not an easy book to read, as Darwin himself admitted.[112] As Huxley grew close to Darwin in the following years he continued to tax Darwin on weaknesses in the natural selection theory until finally they both were mostly satisfied with their explanations. Once convinced, Huxley became Darwin's promoter, often described as "Darwin's Bulldog" having both the energy and good health that Darwin lacked. He edited journals such as *Natural History Review* and acquired administrative positions as a means of promoting Darwin's ideas as well as science in general. He remained always an advocate for scientifically proven truth.[113] During 1859 he carried on through a personal tragedy similar to Darwin's—losing a child. The difference for Huxley was that he had already intellectually rejected traditional Christianity, although he followed the forms of it in giving the child a Christian burial.

Huxley's writings at first are concerned with palaeontology and biology and later encompassed philosophy and theology. Part of his promotion of Darwinism became an engagement with theology on its own terms.

Challenging Christianity—Sensation

Huxley was anti-Paleyan but, unlike Darwin, was not steeped in Paley's theology or committed to the responsibilities of being part of the establishment as landed gentry. In addressing the issue of biblical interpretation, Huxley explicitly took apart assumptions and elements of the Augustinian description of inspiration and rejected them. This can be traced in three parts—firstly in his critique of sensation as described by Newton and the ancients. He then refutes the reliability of divine inspiration by referring to inspiration as he had observed it in Pacific Islander culture. Thirdly, he adopted the methodology of developing liberal Protestant scholarship to raise

---

111. Ibid., 104.
112. Desmond, *Huxley: The Devil's Disciple*, 259; Darwin, *Autobiography*, 213.
113. Desmond, *Huxley: The Devil's Disciple*, 280–91.

questions about traditional methods of interpretation. He came finally to a decisively indefinite position on faith matters—logically he was neither able to confirm nor reject faith—he coined the neologism "agnostic" to describe this position.

One essay not often cited is the study "On Sensation and the Unity of Structure of Sensiferous Organs."[114] He delves into the structure and purposes of comparative anatomical structures used for sensation as well as showing a philosophical debt to Hume and Kant. He also made similar comments in his essay on Bishop Berkley and "Has the Frog a Soul?"[115]

The essay on sensation has implications for Huxley's discussion and rejection of inspiration. One comment in this essay is particularly important. Huxley sees no reason "to deny that the mind feels at the finger points, and none to assert that the brain is the sole organ of thought . . . In truth, the theory of sensation, except in one point, is, at the present moment, very much where Hartley, led by a hint of Sir Isaac Newton's, left it."[116] It was noted earlier that Hartley explicitly used Newton's drawing of a parallel between divine agency in humans in inspiration and divine agency in the world to develop his physiology. Huxley reports that Hartley's theory about sensation, the mind and thought had remained at a point which followed Newton's discussion. The Newtonian description of sensation was demonstrated to be basically that of Augustine's and the ancients, which assumed the existence of the soul. Huxley was not ignorant of Leibniz's critique of Newton, but his concern was to argue for materialism and parochially defend his fellow citizen from this foreigner's attack.

> But the doctrine that all the phenomena of nature are resolvable into mechanism is what people have agreed to call "materialism;" and when Locke and Collins maintained that matter may possibly be able to think, and Newton himself could compare infinite space to the sensorium of the Deity, it was not wonderful that the English philosophers should be attacked as they were by Leibniz in the famous letter to the Princess of Wales, which gave rise to his correspondence with Clarke.[117]

---

114. Huxley, "On Sensation and the Unity of Structure of Sensiferous Organs."

115. Huxley, "Bishop Berkley on the Metaphysics of Sensation," 6; Huxley, "Has a Frog a Soul?"

116. Huxley, "On Sensation and the Unity of Structure of Sensiferous Organs," 292–93.

117. Huxley, "Bishop Berkley on the Metaphysics of Sensation," 247. Huxley used the alternative spelling Leibnitz.

In this introduction to his critique of Berkley, he argues that this old metaphysical description cannot stand. Huxley's preference for a perfect natural law over direct divine action is evident in his criticism of the Newtonians.

> Sir Isaac Newton and his followers have also a very odd opinion concerning the work of God. According to their doctrine, God Almighty wants to wind up His watch from time to time; otherwise it would cease to move. He had not, it seems, sufficient foresight to make it a perpetual motion. Nay, the machine of God's making is so imperfect, according to these gentlemen, that He is obliged to clean it now and then by an extraordinary concourse, and even to mend it as a clockmaker mends his work.[118]

Huxley assumed that if God existed, then divine action, design and forethought must all be perfect. While cautious that Leibniz's view might be a "spiteful caricature of Newton's views," Huxley wondered at the fuss about Leibniz's concern that, "Many will have human souls to be material; others make God Himself a corporeal Being."[119]

> At the commencement of the eighteenth-century, the character of speculative thought in England was essentially sceptical, critical, and materialistic. Why such "materialism" should be more inconsistent with the existence of a Deity, the freedom of the will, or the immortality of the soul, or with any actual or possible system of theology, than "idealism," I must declare myself at a loss to divine. But, in the year 1700, all the world appears to have been agreed, Tertullian notwithstanding, that materialism necessarily leads to very dreadful consequences.[120]

"Tertullian notwithstanding" refers to Tertullian's argument for the corporeality of the soul.[121] It is doubtful that Tertullian would have accepted Huxley's equating of corporeality with being material, hence Huxley was rightly cautious. One negative connotation of a material soul was thought to be its mortality. "Mr. Locke and his followers are uncertain, at least, whether the soul be not material and naturally perishable."[122] Huxley revises Berkley's location of sensation in the spiritual arguing that mechanisms of

---

118. Ibid., 248. Huxley adopts Leibniz's criticism of Newton.
119. Ibid., 247, 248.
120. Ibid., 248–49.
121. Tertullian, *de Anima* 7.
122. Huxley, "Bishop Berkley on the Metaphysics of Sensation," 247; Huxley, "Bishop Berkley on the Metaphysics of Sensation," 251–52.

sensation can be located in material anatomy. However, Huxley notes even when this is done Berkley's question still arises at some point. At "the limits of our faculties" Huxley believes the question is unable to be resolved. This essay appears to be written before his coining "agnostic." Huxley's conclusion or rather non-conclusion represents part of the philosophical basis for agnosticism. For Huxley the question of the soul's immorality or even if it existed remained, in cold logic, an open question. Rather than accept any attribute of the soul as a given, Huxley focussed on what was experimentally verifiable. Huxley argued strongly that the soul, if it exists, must be corporeal, have spatial extension and as a material be divisible. These are, though Huxley fails to mention it, similar attributes as assigned to the soul by Tertullian. In "Has the frog a Soul?" he argues thus:

> As the schoolmen supposed the Deity to exist in every ubi, but not in any place, so they imagined the soul of man not to occupy space, but to exist in an indivisible point. Yet whoever considers the structure and appearances of the animal frame, will soon be convinced that the soul is not confined to an indivisible point, but may be present at one and the same time, if not in all parts of the body, when the nerves are formed, yet, at least at their origin, i.e., it must be at least diffused along a great part of the brain and spinal marrow. Nay, while in man the brain is the principal seat of the soul, where it most eminently displays its powers, it seems to exist or act so equally through the whole bodies.[123]

Huxley determines the anatomical location of the soul by its supposed function. The notion of the soul he works with is the Augustinian description in Newton's application of it. "Some of the greatest philosophers" obviously includes Newton.

> It is not, therefore, altogether without reason, that some of the greatest philosophers of the last and present age supposed the soul to be extended....
>
> As the Deity is everywhere present, and, in the infinitely distant part of space, actuates at the same time a vast variety of different systems without any inconsistency with his unity or indivisibility; so may not the souls of animals be present everywhere in their bodies, actuating and enlivening at the same time with all their different members? Nay, further, when the fibres and threads connecting some of these parts are divided, may

---

123. Huxley, "Has a Frog a Soul?"

not the soul still act in the separated parts, and yet be only one mind?[124]

Huxley's reference to philosophers here cites More, Newton and Clarke. The material nature of the alleged soul requires its extension through the body as demonstrated by the distribution of its functions. Huxley argues with reference to vivisection:

> A frog's head is cut off so that the section passes between the medulla oblongata and the rest of the brain. The actions performed by the head and by the trunk will be equally purposive, and equally show that there is a something in each half which possesses the power of adapting means to ends in a manner which is as deserving as the epithet "rational" in the one case as in the other. The separated head and trunk may be sent a hundred miles in opposite directions, and at the end of the journey each will be as purposive in its actions as before. In this case, two alternatives present themselves,—either the soul exists in both cord and brain, or it exists in only one of them.[125]

Because functions normally attributed to the soul are distributed anatomically, he thus argues that if the soul exists it must also be distributed and, having the attributes of matter, cannot therefore be metaphysical. "I am unable to see in what respect the soul of the frog differs from matter." Despite his rejection of many aspects of the Augustinian description of inspiration, the spatial bodily extension of the soul is compatible with Tertullian's description described earlier. Huxley had an advantage over Newton. What Newton described as connections, nerve fibres, were beginning to reveal their function under the microscope and in the biology laboratory. While Huxley recognized that there were limitations on what previous theory could determine, the fact was that:

> The sense organ is not a mere passage by which the "*tenuia simulacra rerum,*" or the "intentional species" cast off by objects, or the "forms of sensible things," pass straight to the mind; on the contrary, it stands as a firm and impervious barrier, through which no material particle of the world without can make its way to the world within . . . Interconnection of all these three

---

124. Ibid.

125. Ibid. The question of ethics and vivisection was itself a controversial topic contributed to by Darwin and Huxley.

structures, the epithelium of the sensory organ, the nerve fibres, and the sensorium, are essential conditions of ordinary sensation.[126]

Luigi Galvini's discovery of bioelectricity and related animal movement only occurred 8 years before in 1862 as a serendipitous discovery involving dead frogs.[127] Huxley and his contemporaries merely described the anatomical structure of nerves. Charles Sherrington,[128] a generation after Huxley, described the electrical nature and role of nerves in detail. The connection of senses and the ability of the mind to direct muscular movement became no longer the mystery ascribable to the metaphysical or spiritual interaction of the soul in the sensorium that they were for Newton and Augustine. *Ekstasis,* as previously understood, would in this newly discovered network of nerves, effectively require a displacement of the physical brain or at least a total temporary rewiring.

> On the contrary, the inner ends of the olfactory cells are connected with nerve fibres, and these nerve fibres, passing into the cavity of the skull, at length end in an element of the brain, the olfactory sensorium. It is certain that the integrity of each, and the physical interconnection of all these three structures, the epithelium of the sensory organ, the nerve fibres, and the sensorium, are essential conditions of ordinary sensation.[129]

As Huxley pointed out such a rewiring of nerves is clearly impossible. Thus *ekstasis* in Augustinian-Aristotelian terms is also impossible. How could the agency of the Holy Spirit in humans necessarily require *ekstasis*? Something is wrong. Either the soul does not exist as Huxley suggested but could not prove, or God does not work by the Holy Spirit, or alternatively what Huxley failed to explore is that the explanation of the agency of the Holy Spirit is inadequate.

Challenging Christianity—Inspiration

Huxley shares a perception common in his time that inspiration which came by divine agency ought to be infallible. He shared the preconception but did

---

126. Huxley, "On Sensation and the Unity of Structure of Sensiferous Organs," 300–301.
127. "The Wordsworth Dictionary of Biography."
128. Ibid.
129. Huxley, "On Sensation and the Unity of Structure of Sensiferous Organs," 300–301.

not believe it. In his provocative "The Evolution of Theology," he tackles the question of inspiration directly. At first, he traces a deprecating version of biblical history citing unusual accounts of inspiration in the Old Testament to establish a justification for a simplified and standardized method of analysis. He then relates these to phenomena he had observed elsewhere. His assumption is that if some manner of direct divine or spiritual communication exists, then it must be a common human faculty. Therefore, he assumed, generic study by comparison is possible. He applied a scientific methodology to the sacred text merely taking events recorded as data. While he believed himself to be objective, his method was theory laden with a set of assumptions loosely derived from early higher biblical criticism. In keeping with what Frei has described as a nineteenth-century trend in biblical studies, Huxley uses the narrative as a place to mine nuggets of data thereby eclipsing the story.[130]

> I need hardly say that I depend upon authoritative biblical critics, whenever a question of interpretation of the text arises. As Reuss appears to me to be one of the most learned, acute, and fair-minded of those whose works I have studied, I have made most use of the commentary and dissertations in his splendid French edition of the Bible. But I have also had recourse to the works of Dillman, Kalisch, Kuenen, Thenius, Tuch, and others, in cases in which another opinion seemed desirable.[131]

Huxley's definition of fair mindedness would not have been shared by his English contemporaries. Kalisch was a liberal and politically radical Rabbi;[132] Dillman was an early higher critic; Kuenen, Thenius and Tuch were rationalists;[133] and Reuss initially proposed the documentary hypothesis later elaborated by Wellhausen.[134]

Huxley assumed an evolutionary progression in theological development moving from more primitive animistic spiritualism to a more refined monotheism. His first example attempted to draw common patterns in the inspiration of the witch of Endor, Samuel, and the kings Saul and David.

> The wise woman of Endor was believed by others, and, I have little doubt, believed herself, to be able to "bring up" whom she would from Sheol, and to be inspired, whether in virtue of

---

130. Frei, *The Eclipse of Biblical Narrative*, 51–65.
131. Huxley, "The Evolution of Theology," 4:294n3.
132. "The Columbia Encyclopaedia."
133. Aherne, "Commentaries on the Bible," 4.
134. Reid, "Biblical Criticism (Higher)," 4.

actual possession by the evoked *Elohim*, or otherwise, with a knowledge of hidden things, I am unable to see that Saul's servant took any really different view of Samuel's powers, though he may have believed that he obtained them by the grace of the higher *Elohim*.[135]

Huxley here assumes that these experiences are similar in form without reading the narrative's interpretation of the events that unfold. Huxley cites the example of Saul apparently conducting the process of divination in 1 Samuel 14 and claims that David seems to do the same. Huxley argues for a kind of egalitarianism of inspiration among "professionals" like the Endor seer (not the pejorative "witch") and Samuel and then extending this to the general public. "Although particular persons adopted the profession of media between men and *Elohim*, there was no limitation of the power . . . to any special class of the population. Jahveh himself thus appears to all sorts of persons, non-Israelites as well as Israelites."[136] He finds significant the accounts of involuntary inspiration such as visitations in dreams and of people without their volition. "Again, the *Elohim* possess, or inspire, people against their will, as in the case of Saul and Saul's messengers, and then these people prophesy—that is to say, 'rave'—and exhibit the ungoverned gestures attributed by a later age to possession by malignant spirits."[137] Huxley saw direct parallels between Saul's *ekstasis* and his own observations. He does not doubt the biblical accounts as he had observed what he understood to be similar phenomena. "Apart from other evidence to be adduced by and by, the history of ancient demonology and of modern revivalism does not permit me to doubt that the accounts of these phenomena given in the history of Saul may be perfectly historical."[138]

At this point in his argument, he is content to reduce the processes to some common elements upon which he can draw parallels. What he did not accept in any way, however, were the traditional Christian or Jewish interpretations of the events. Later, he was to describe all inspiration as merely a psychological oddity. The pivotal argument in the "Evolution of Theology" is made by relating his construct of inspiration by *Elohim* "possession" to the understanding of Pacific Islander religion he gained while travelling with the *Rattlesnake* in 1848. Firstly, he refers to his direct experience in the Torres Straits. "This scene made an impression upon me which is not yet effaced. It left no question on my mind of the sincerity of the strange

---

135. Huxley, "The Evolution of Theology," 303.
136. Ibid., 305, 306.
137. Ibid., 306.
138. Ibid.

ghost theory of these savages, and of the influence which their belief has on their practical life."[139] He then expands the parallel to pre-Christian Tongan religion quoting Mariner's work.

> Moreover, the *Atuas* were believed to visit particular persons,— their own priests in the case of the higher gods, but apparently anybody in that of the lower,—and to inspire them by a process which was conceived to involve the actual residence of the god, for the time being, in the person inspired, who was thus rendered capable of prophesying. For the Tongan, therefore, inspiration indubitably was possession.[140]

Huxley belabored the similarity between this and the Old Testament accounts as he has interpreted them. Firstly, he identifies possession and then the notion of indwelling: "As soon as they are all seated the priest is considered as inspired, the god being supposed to exist within him from that moment."[141] Having drawn a parallel, Huxley refutes infallibility and then proposes a natural explanation. As the accounts are similar but the resulting "revelations" are incompatible, therefore no "revelation" can be reliably true. The materialist explanation he gives is as follows:

> The phenomena thus described, in language which, to any one who is familiar with the manifestations of abnormal mental states among ourselves, bears the stamp of fidelity, furnish a most instructive commentary upon the story of the wise woman of Endor. As in the latter, we have the possession by the spirit or soul (*Atua, Elohim*), the strange voice, the speaking in the first person.[142]

Huxley claimed such accounts of ancient times are common aberrations which the "civilised" person must outgrow with the use of reason and science. Huxley implied that if these phenomena are considered as aberrations the resulting inspired instructions cannot logically be given ultimate authority, but must be subject to reason which is more trustworthy. In his twice published essay "Witness to the Miraculous" he explicitly stated this rejection of revelation in favor of reason. He daringly goes as far as to say that dependence on the miraculous for proof of religion goes counter to the

---

139. Ibid., 317–18.
140. Ibid., 323–24. *Auta* is Tongan for spirit.
141. Ibid., 325.
142. Ibid.

intent of the original writers, "that they would regard the demand for it as a kind of blasphemy."[143]

The growth of complex forms of religion is something he believed was a sign of intellectual progress. Huxley argued that this had already occurred early in history as in the allegory of Philo and Zeno. Allegory, however, became problematic through overuse. Some alternative method for resolving problems caused through complexity of belief was needed. "This mighty 'two-handed engine at the door' of the theologian is warranted to make a speedy end of any and every moral or intellectual difficulty."[144]

If intellectual difficulties are to be considered as part of the natural development of a religion they might in turn be better resolved in the light of the some general law rather than by what he saw as their avoidance through the over use of allegory. Moral law cannot come about, he argued, by the recognition of revelation. The refined religious intellect should see no place for this miraculous intervention. Huxley firmly believed that moral law does exist, but can only be conceived as one grasps rationally the shape of a higher general law of nature independent of theology.

> It is my conviction that, with the spread of true scientific culture, whatever may be the medium, historical, philological, philosophical, or physical, through which that culture is conveyed, and with its necessary concomitant, a constant elevation of the standard of veracity, the end of the evolution of theology will be like its beginning–it will cease to have any relation to ethics.[145]

Huxley's motivation for explaining away inspiration was to find an alternative ground for ethics. The miraculous could not be used as a justification for ethics as much as it could not be used to justify God's existence. Huxley was aware of a need to interpret Scripture and nature in a manner different to the fourfold method of the Middle Ages in which allegory played an important part. However, in doing so, he also rejected the two books metaphor. How could either the bible or nature be considered books of revelation when there could be no authoritative inspiration upon which the two books notion is so firmly based?[146]

---

143. Huxley, "Witness to the Miraculous," 5:190–91.
144. Huxley, "The Evolution of Theology," 366.
145. Ibid., 372.
146. Huxley, "Agnosticism and Christianity," 8–9.

Challenging Christianity—Liberal Protestantism

Whereas Darwin's rejection was of the Paleyan construction of Christianity, Huxley went a step further and rejected the assumptions underpinning the Paleyan consensus. In addition to rejecting immutability, Huxley had also rejected the old theories of sensation and inspiration. Along with evidence from geology, palaeontology and from Darwin's theories, Huxley began a critique of traditional belief using what was then state-of-the-art biblical scholarship from the developing liberal Protestantism. He cites Strauss as forcing scientific theology to take into account the development of the Gospel narrative, and Baur for raising to prominence the divergence in Nazarene and Pauline tendencies in the primitive Church.[147] Huxley repeatedly draws a distinction between the earliest Nazarenes and Christians with reference to Baur's theories regarding theological debates prior to the fourth century.

Huxley demanded application of a consistent scientific methodology including a constancy of interpretation of biblical vocabulary similar to Wellhausen's documentary theory. Although he did not offer an opinion on the documentary hypothesis, it would be consistent with his emphasis on scientific consistency that he would have dismissed the theory that four traditions had been edited into one text as a feeble attempt to systematize inconsistency. He did not use German scholarship uncritically. While he favoured the developing liberal Protestant approach he flavoured that support with the scientific scepticism he so valued. His primary attack on biblical methodology countered the authority of the Church over that of reason. His bias against the hegemony of the Paleyan Anglican elite remained with him all his life.

> Nothing can thus be clearer than that the Church of England places the Bible above "the Church," and gives it an authority which is independent of the Church. It speaks of a fallible Church and an infallible Bible. It represents the Church as being to the Scripture what the High Court of Justice is to the statutes.[148]

It is reason, not an institution, Huxley argued, which should be the arbiter of interpretation. Huxley was so thoroughly scientific that he went as far as to describe belief in immortality as a sin in the absence of evidence,[149] and was dubbed the "Apostle Paul of the new teaching," by the *Daily News*.[150]

147. Ibid., 361–62.
148. Huxley, "An Apologetic Irenicon in the Fortnightly Review," 1–2.
149. Desmond, *Huxley: The Devil's Disciple*, 288.
150. Desmond, *Huxley: Evolution's High Priest*, xiii.

There is little doubt that the general public understood that Huxley and the Darwinists were controversially promoting a different religion. The same Anderson whose initial article prompted Huxley's 1892 article "The Bible and Modern Criticism" in the *Times* responded:

> If facts be adduced to prove the Bible false, I shall give it up and cease to be a Christian. But practical men and men of common sense care little for mere theories. In common with so many other Christians I regard the Darwinian theory of evolution as being, within strictly defined limits, a reasonable hypothesis.[151]

Anderson demonstrated the intensity of feeling, and the all-or-nothing belief in Scriptural infallibility. His simultaneous acceptance of a form of Darwinian Theory was also not unusual. He continued; "Nor am I abashed at incurring Mr. Huxley's contempt for the statement I made that the Scriptures are, as Lord Bacon phrased it, 'of the nature of their author' and have a deep spiritual meaning and a 'hidden harmony' far beneath the surface strata in which the critics ply their tools."[152] Huxley's naturalistic explanations found no support here. Nevertheless, Huxley summarized his approach in his response to Anderson on February 11, 1892.

> Science, like nature, may be expelled, but not even the ecclesiastical work can keep her from coming back. By those who profess to be guided by anything better than instinct, the credentials of the "infallible" authority must be submitted to reason. The so-called sacrifice of private judgment is in fact the apotheosis of private judgment.[153]

What had been generally accepted explanations of the Christian faith used in the early nineteenth century did not survive with the same wide support by the century's end. Huxley called for and received significant support for revising key parts of the Christian faith such as perfect divine action and arguing for the existence of God from the miraculous.

This chapter began by noting that a parallel had been drawn between divine agency in humans as described in *ekstasis* inspiration and divine agency in the world complimenting early eighteenth-century understandings of the notion of the two books and divine perfections. The validity of this understanding of divine agency in the world was understood to depend on divine perfection being reflected in the laws and design of nature. This

---

151. Anderson quoted in Huxley, "An Apologetic Irenicon in the Fortnightly Review," 15.
152. Ibid.
153. Ibid.

was thought to mean that: there was a cause for everything; God acts with a motive and acts for the best and; the world was the best of possible worlds and demonstrated God's purposes. Also, this perfection in nature was put in place by God without error just as the agency of God's inspiration achieved this for Scripture. Darwin laid aside each aspect of purposeful perfected divine action in nature. Huxley further attacked the internal logic of the Augustinian inspiration by discounting both the existence of the soul and the authority of revelation by inspiration. The question remains, why does Christianity persist? Is it cultural inertia or is it, as will be argued later, that perfect divine action in the world and the existence of the soul following the logic of this form of divine agency do not occupy the foundational role in Christian faith which its critics assumed it did? With a new description of divine agency which is grounded in who God has revealed God's self to be it will be argued that it is possible to overcome Darwin and Huxley's objections.

The Legacy—the Shape of the Stumbling Block to the Dialogue between Theology and Science

Paley used the perfect adaptation of creatures to their environments as evidence of the perfect hand of the creator at work. To suggest as Darwin did that species have adapted over time to fit their environment negated the Paleyan argument from design and, further, Huxley removed the basis for the traditional understanding of the soul as a metaphysical entity. What has been argued is that the problem in its simplest form is that an unresolved tension arises if divine agency is necessarily linked to non-christological perfect-being theology and the existence of the soul. This particular issue can be resolved if divine agency can be theologically re-expressed without this dependence. However such resolution has been complicated by conflation with myths and misleading generalisations. This has resulted in an apparently larger stumbling block to the dialogue between theology and science in which the choices have become ostensibly limited to either revising theology or limiting God's perfection or rejecting the faith. What will be argued is that this limited range of choices derives from a logical fallacy which may be overcome by revising divine agency in terms of the incarnation. It is necessary to identify these conflating myths and misleading generalisations in order to show that the specific tension relating to divine agency can be resolved by revision of its theological description.

Apart from a continued obscurantist reaction, which refused and still refuses to even consider the new science, initial reactions to Darwinian thought have been broadly three-fold. Firstly, some like Darwin and Huxley rejected traditional Christianity moving to a place of decided uncertainty, i.e. agnosticism. Secondly, others rejected faith altogether and moved to atheism. Thirdly, yet others argued for the consistency of truth either actively embracing newly developing liberal Protestantism or revising specific areas of biblical interpretation. In the late nineteenth century the established presuppositions were not re-examined to determine whether it was possible to pose additional choices for theological response. The existence of myths and generalisations has had two effects which have not been helpful. They have oversimplified the complexity and diversity of reactions to Darwin and Huxley implying that tensions arising from their work are inseparable from large concerns. Additionally, they have led to the neglect of detail and consideration of carefully argued historical opinion which might suggest that tensions might be dealt with in little-by-little rather than across-the-board fashion.

False Myths and Misleading Generalizations

There are three myths and generalizations that need to be set aside: Firstly, that Darwinian Theory defeated all comers; Secondly, that in the warfare between science and religion, science has triumphed, and; Thirdly, all opposition to Darwin and Huxley was non-scientific. What is argued is that by removing these myths the question of divine agency can be demonstrated to be one of a number of possible sources of unresolved tension in the dialogue which might be valuably dealt with independently of other questions.

Darwinism Defeated All Comers

The least accurate myth has been that Darwinian evolutionary theory has simply triumphed over all comers. While Huxley would have everyone believe that all "liberal reconcilers of Christianity and evolution could be nothing but an 'army' bent on blending scientific truth with theological error,"[154] many so-called "liberal reconcilers" have attempted just this. Contrary to Huxley and the myth-making heroic and Whig histories of science, Peter Bowler has clarified by careful historiography that Darwin was not the all-conquering voice of reason in the wilderness of ignorance, and that Darwinism was not the unstoppable juggernaut of scientific progress.[155]

---

154. Moore, *The Post Darwinian Controversies*, 217.
155. Bowler, *The Eclipse of Darwinism*.

No simple model can describe the legacy of Darwinian Theory at the turn of the twentieth century. Bowler argues that

> the apparent lack of interaction between science and religion in the early twentieth-century is an artefact of historians' neglect. . . . issues were not as dead as the lack of historical emphasis might imply. Old topics such as the implications of evolutionism still attracted the attention of scientists and religious leaders and could still generate headlines in the popular press.[156]

The fact that the lack is merely apparent is highlighted by Morton's long bibliography of primary sources related to evolution's effect on the literary imagination in England of this period.[157] The range and complexity of issues addressed led to some markedly divergent reactions, such as the Scopes trial in the United States and Chadwick's observation that "in 1900 men talked as though the conflict [with science] was over."[158] As Bowler comments, "the fact that two English-speaking countries could experience so different a chain of events during the same period raises the prospect"[159] that there are issues of wider significance involved. Darwinian natural selection waxed and waned in academic favour, but so also did optimism about the possibility of a resolving tensions between theology and science.

Science Defeated Religion

Another persistent myth describing the relationship between theology and science since the late nineteenth century has been that of conflict or warfare. While Huxley promoted a conflict model, his conflict was most strictly between science and established ecclesial authority. As Bowler notes,

> For all his professed lack of belief in a personal God, Huxley continued to suppose that evolution was ultimately a purposeful process. And, unlike some more materialistic scientists, he accepted that religious experiences had a genuine value for human life.[160]

---

156. Bowler, *Reconciling Science and Religion: The Debate in Early Twentieth Century Britain*, 2.

157. Morton, "Darwinism and the Victorian Literary Imagination"; Morton, *The Vital Science*.

158. Bowler, *Reconciling Science and Religion*, 1–2; Chadwick, *Victorian Church*, 2:35.

159. Bowler, *Reconciling Science and Religion*, 3–4.

160. Ibid., 14.

The broadening of the conflict myth probably owes as much to the widely read works of Draper and White[161] promoting the warfare of science and religion as to any other sources. There is general agreement that both Draper and White are misleading and ideologically driven. Harrison refutes their base metaphor indicating that prior to the mid-nineteenth-century "religion" referred to personal piety and "science" to knowledge or wisdom.[162] Brooke and Cantor re-examine the persistent false myths, in the light of actual events and documentation.[163] Those myths are: 1. Columbus against the "flat earthers," which was actually a disagreement over the actual diameter of the world; 2. Galileo and the inquisition, where Galileo was actually jailed for insulting the pope who was a former sponsor of his work; 3. Darwin's loss of faith as a result of his theory, which actually occurred as shown here after the death of his daughter Annie.

Moore, for example, notes according to the warfare myth, the conservative Hodge is usually described as an obscurantist bibliolater who simply equated Darwinism with atheism rather than being one of the first in the evangelical tradition to give a deep theological analysis of Darwin's theory. Moore describes Hodge as among Darwin's more discerning critics across the Atlantic and Hodge's three volume Systemic Theology was a masterful attempt to adapt theology to the methodology of Newtonian science. Hodge's disagreement with Darwin was on matters of theory rather than matters of fact.[164] This not to say that Hodge was impressed with the way natural selection lent itself to anti-spiritual materialism. Hodge criticized Darwin scientifically along Baconian lines and did not intrude the bible into his discussions.[165]

To further illustrate the complexities which destroy the myth, the conservative theologian Warfield was originally quite open to Darwinism. He corresponded regularly with Asa Grey who was both a Darwinian and a devoted Congregationalist. Like his mentor Hodge, Warfield saw a place for evolution in a conservative reformed theology. Warfield cautiously rejected Darwinian natural selection in the early twentieth century (1908),[166] at a

---

161. White, *A History of the Warfare of Science with Theology in Christendom*; Draper, *History of the Conflict between Religion and Science*.

162. Harrison, "The Book of Nature Metaphor and Early Modern Science."

163. Brooke and Cantor, *Reconstructing Nature: The Engagement of Science and Religion*.

164. Moore, *The Post Darwinian Controversies*, 193, 204.

165. Ibid., 211.

166. Warfield, "Review of George Paulin, No Struggle for Existence: No Natural Selection. A Critical Examination of the Fundamental Principles of the Darwinian Theory," in Noll and Livingstone, *Warfield*, 252–56.

time when natural selection had for a period generally fallen out of favor in the scientific community as an explanatory mechanism for species change. This period extended from 1905 to 1920. Darwinian natural selection gained a new academic lease of life in the 1930's when point mutation was added to natural selection theory as a possible mechanism for explaining species change. The alternatives will be discussed later in this section.[167]

All Opposition to Darwinian Is Non-scientific

Darwin himself had doubts about the universality of his theory. The spectacular display of peacock's plumage gave him considerable discomfort as it seemed to Darwin to serve no evolutionary benefit. This discomfort surfaced when John Ruskin gave Darwin some studies on peacock feathers following a dinner party. Ironically recent research indicates that the better and larger the display of a Peacocks tail, the healthier and more fertile the bird and demonstrably the more successful that bird's offspring.[168] Similarly, there were a number of reservations expressed about Darwinism by scientists as well as theologians regarding its structure, applicability and logic.

James Moore highlights the range of differing opinions and reservations raised as people responded both theologically and scientifically to Darwin.[169] These reservations included concern about Darwinism's tendency to materialism, both in its effect on ethics as well as lending itself to a mechanism of the mind which could exclude the Spirit.[170] However, the reduction of emotional and behavioural characteristics of humans to biochemistry and neurology can only be automatically threatening when Christianity is assumed to depend on a metaphysical soul to which these attributes are supposed to belong. There was concern about what many saw, and still see, as the most important feature of modern Darwinism, i.e. its "elimination of any need to see change as a goal-directed."[171] Eliminating purpose has been both theologically and scientifically problematic. It is theologically problematic for the current discussion of divine agency in the world, because it denies the reality of such agency. It has been scientifically problematic for those scientists who do not think that natural selection alone can explain the upward progression of the natural order.

167. Bowler, *Reconciling Science and Religion*, 3.
168. Desmond, *Huxley: The Devil's Disciple*, 638; Dawkins, *The Selfish Gene*, 309–13.
169. Moore, *The Post Darwinian Controversies*, 193–298.
170. Bowler, *Reconciling Science and Religion*, 160. Bowler, "Evolution and the Eucharist," 453–67; Brooke, *Science and Religion: Some Historical Perspectives*, 281.
171. Bowler, *The Non-Darwinian Revolution: Reinterpreting a Historical Myth*, 176.

Natural selection as a mechanism for change had competition. Bowler has identified four broadly competing evolutionary theories in scientific discussions after the turn of the twentieth century. The first was Darwinian natural selection. The second was Theistic evolution in which God ordains species change (a logical heir of Sedgwick and Owen). The third was a kind of Neo-Lamarckism derived from the late nineteenth-century theory in which characteristics acquired during life are passed onto a creature's offspring. The fourth, orthogenesis, described progressive development of species by forces originating within the organisms.[172] Each of these competing theories has enjoyed varying degrees of success and support from the beginning of the twentieth century to the present day. The December 1997 *Quarterly Review of Biology* contained a series of articles on evolution and theology which represent a number of these four theories. It also includes articles by John Paul II, Michael Ruse and Richard Dawkins.

Criticism of Darwinism by Christian thinkers where it occurred often was on scientific rather than theological grounds. In one case Huxley's student George Mirivart[173] ceased being a Darwinian when he came to believe that natural processes alone could not explain upward development of life.[174] In response, Huxley and the other members of the X virtually ostracized him.

Another example is the dispute in the USA between two Christian pioneers of the new sciences, Asa Gray and Louis Agassiz.[175] Whereas, Gray worked with Darwin on evolutionary research in plants, Agassiz as a young man in Europe had developed glaciation theory which had been welcomed by Lyell and contributed to by Darwin. He had moved to the USA by the time of the publication of *Origin*. A dispute between Agassiz and Gray regarding natural selection is described by Croce. Agassiz's reasons for his rejection of Darwin were scientific and methodological. For Agassiz new species appeared only at divine instigation.[176] Agassiz adhered to the notion of fixity of species supported by the data. What was allowed was structural flexibility within a species but variations tended to return to form. The development of new species Agassiz still saw as the providential preconceived plan of creation. Gray disagreed. In collaboration with another conserva-

---

172. Bowler, *The Eclipse of Darwinism*, 7. A later theological derivative of the last is Teilhard de Chardin's evolution toward the Omega point. Teilhard de Chardin, *The Phenomenon of Man*, 235–300.

173. Chair of Zoology at St Mary's College London.

174. Bowler, *Charles Darwin: The Man and His Influence*, 163.

175. Croce, "Probabilistic Darwinism: Louis Agassiz Vs Asa Gray on Science, Religion and Certainty."

176. Moore, *The Post Darwinian Controversies*, 208–10.

tive, Wright, he held that Darwinism was not necessarily inimical to Paley's argument rightly understood.[177] Ongoing correspondence between Gray and Darwin highlights Darwin's lack of certainty about the philosophical implications of his theory and his tendency to continue to use the language of purpose or providence when writing about change.

Wright himself is another complex example defying simple categorization. Wright, concerned about the increasing influence of liberalism, was actually one of the contributors to *The Fundamentals*, becoming a prominent theological conservative.[178] *The Fundamentals* led to the coining of the contemporary term fundamentalism. His is a complicated academic career whose detail does not fit the simplistic "warfare" myth. Nevertheless, many commentators from the 1960's to the 1980's typically regard the early Fundamentalists in negative and dismissive terms. Moore, for example, describes Wright's involvement in Fundamentalism as a sad end to his career. Gillespie and Rupke are similarly hostile.[179] Brooke and Cantor, disagreeing with Gillespie and Rupke, argue that Wright like others derived his position from a commitment a scientific tradition of non-speculative thought. Brooke and Cantor, provocatively link present day creationist and biblical geologist Fairholme with Wright suggesting both charge their, "contemporary geologists with indulging 'the very excesses of hypothesis.'"[180] Brooke and Cantor suggest that the accusation of the scientific sin of excessive speculation directed at Wright and Fairholme might as easily be pointed back at the accusers.

Theistic evolution, Lamarckism and orthogenesis are attempts to account for progress or improvement in the development of natural organisms which do not seem to arise by chance. This hesitancy about the language of purposeful design has never completely removed from the science. For example, as recently as 1992 while commenting on the varying fortunes of Haeckel's ontology and phytology during the late nineteenth and early twentieth centuries, Stephen Jay Gould has argued for a limited return using these in describing progress in evolution.[181] It is not clear in that paper, however, whether he is arguing for a theistic, a Lamarckian, an orthogenic approach or is merely leaving the question open.

---

177. Bowler, *Charles Darwin: The Man and His Influence*, 160.

178. Moore, *The Post Darwinian Controversies*, 280, 290, 296.

179. Gillespie, *Genesis and Geology*, 152; Rupke, *The Great Chain of History*, 218.

180. Brooke and Cantor, *Reconstructing Nature: The Engagement of Science and Religion*, 57–59.

181. Gould, "Ontogeny and Phylogeny—Revisited and Reunited," 275–79.

Rather than being non-scientific in their responses, Barth suggests that the problem among theologians was their too ready acceptance of the science. Barth was concerned about the enthusiasm for promulgating evolution even among those who should have critically examined it. Speaking of the late nineteenth century, Barth describes this as

> the period when the "descent theory," or more precisely "transmutation theory," founded by Jean-Baptiste Lamarck, Lorenz Oken and Charles Darwin and represented especially by Ernst Haeckel, had reached its height. The theologian was confronted by the theory—alarming because [it was] so enthusiastically expounded and supported by such a wealth of illustration.[182]

Barth protested that theologians abdicated questions regarding the ontological purpose to science. The place of humanity in the order of creation was consequently no longer taken as a given even in theological discussion. "What was now doubted and contested as a result of a new emphasis in modern science and the outlook to which it gave birth, was the idea of the special position of man in the universe."[183] Barth referred to Haeckel's phylogenesis and embryology reflecting his influence in the development of evolutionary theory in German language thought. Haeckel's work had spiritual overtones than the more materialistic turn of Darwin and Huxley in the English speaking world. In Barth's analysis, humans had become within evolutionary theory in Germany a spiritual being whose separation from the rest of the created order depended on an assumed character of "good" overcoming "evil." Noting Hegel's use of the theory, Barth states, "Yet for him, too, 'it is a basic requirement of piety, to be retained at all costs, that the human mind is to be regarded as far above the rest of creation, and of quite a different order from that of minerals, plants and animals.'"[184] This abdication to the philosophy and terminology of nineteenth-century science had made theological critique difficult. "It would have been hard for the nineteenth century to deal with the Darwinians if it had not been for the good fortune that even among scientists there were more or less intelligent and resolute anti-Darwinians to whom the theological apologists could refer."[185] Barth was blunt in his criticism of the nineteenth-century theological response to Darwinism as immodestly adopting both the theories and the values of the science. "First, they shared the arrogance of the Darwinians to the extent that they increasingly accepted Darwinian theories as a secure basis for all

---

182. CD III/2:80.
183. Ibid., 79–80.
184. Ibid., 81.
185. Ibid., 88.

further progress. And second, they thought it their duty to complete and transcend these theories by opposing to them the further dogma of man as an intellectual and cultural being."[186] His criticism is that his predecessors abdicated the essential nature of theological discourse and assumed more from science than it could give. This, Barth argued, lost the essential theological truth of the ontology of humanity, that: "Man exists in the fact that what he is told by God is the truth. He exists in this truth and not apart from it."[187] Barth in his *Church Dogmatics* left open the question of an appropriate theological response to Darwinian Theory.

Darwin's Legacy: A Clearer Picture May Lead to Resolution of Underlying Unresolved Issues

The promulgation of these false myths has suggested that the apparent limitation of choices (to either revising theology or limiting God's perfection or rejecting the faith) can only be considered as part of larger issues. Theological engagement appears to come out of silence and defeat and scientific ignorance. Whereas, the actual historical complexity suggests that the tensions in the dialogue arise from a number of separate issues of which the question of divine agency is but one issue, albeit an important one. Rather than responding non-scientifically, theologians have too readily responded using science as the foundation of their analysis, thus compounding non-theologically those problems that have been argued to have arisen using non-christological generic theology in developing divine agency.

These three misleading notions, that Darwinian Theory trumped all opposition, that science "defeated" religion and all opposition has been non-scientific have contributed to the pessimism Bowler expressed in wondering if expecting the underlying issues to ever be resolved may be futile. These misleading notions have been used to support wider ideological agendas; the cases of Huxley, Draper and White are examples. Where this is the case or the misleading notions have remained unacknowledged the result has been to obscure detail in the tension which might be resolved or preclude criticism as "non-scientific." Bowler describes past attempts to reach resolution:

> There have been three major episodes in the twentieth-century when interest in the possibility of constructing a reconciliation between science and religion has flared. The first occurred in the

---

186. Ibid.
187. Ibid., 152.

early decades of the century and forms the theme of this book. The second wave . . . began in the aftermath of World War II and lasted into the 1960's. The third seems to have arisen quite recently. . . . The tensions of the Victorian era have thus been sustained throughout the twentieth-century, each episode of challenge being followed by one of attempted reconciliation.[188]

Theological responses to Darwinism have varied considerably, and dialogue has never reached conclusion. Bowler points out that in the generations since Darwin, there have been repeated patterns of inconclusive discussion between theology and science as interest has waxed and waned from generation to generation. Indeed, some of the aspects and trends in the current dialogue between theology and science bear more than passing resemblance to other earlier dialogues in the early twentieth century.

With the obscuring of detail needed to identify individual issues which might be resolved, the dialogue has been faced with alternatives which have received little enthusiasm. Bowler describes the failure of the 1930's attempt at reconciliation of science and religion. "It was the theologians, at least as much as the scientists, who turned away from the proposed reconciliation."[189]

> Not everyone accepted the proposed synthesis of science and religion, of course. Secularists were still active and wanted to use science as a weapon against all forms of what they regarded as superstition. . . . The proposed synthesis required the modification, if not the actual suppression, of many aspects of traditional Christian belief, reducing religion to a more generalized theism. The synthesis thus depended on a degree of theological liberalism that many orthodox Christians regarded as a complete betrayal, and their fears were highlighted by the fact that it was endorsed by some openly non-Christian writers such as Shaw.[190]

While unresolved underlying issues between theology and science remain, it is likely that patterns of argument like these examples will be repeated. At the turn of the twenty-first century, Baker noted that some responses to Darwinian theoretical problems were a return to a form of Paleyanism.[191] Baker is correct in stating that any form of Paleyan theology is inadequate and that some new form of narrative is needed to describe the natural world. Nearly a century after Barth, in *Nein!* criticized the inad-

---

188. Ibid., 4.
189. Ibid., 286.
190. Ibid., 3.
191. Baker, "Theology and the Crisis in Darwinism," 183–215.

equacy of Paley in Brunner's reformulation of natural theology, the debate remains alive in both theology and in science.[192]

It is argued by identifying and removing misleading notions that a clearer understanding of the individual issues leading to tensions in the dialogue may be achieved and that repetition of the patterns of past dialogue might be overcome. This book seeks to show that such resolution is possible for one issue, divine agency, whose development has been shown to be linked with the early modern understanding of the divine perfections. With this linkage, the understanding of divine agency developed by the nineteenth century into an unworkable Paleyanism. Perfection and the existence of the soul had come to be included among the assumed preconditions of the Christian faith. Darwin and Huxley found these "assumptions" wanting and used these assumption among others as reasons for rejecting traditional Christianity. What remains unknown is what might have been different in the development of scientific theory if an alternative description of divine agency in humans had been applied. If it is possible to redescribe divine agency so as to avoid the necessity of a link with perfect-being theology, then it is possible to resolve one of the underlying issues leading to tensions in the dialogue between theology and science.

A Logical Fallacy

There is a formal logical fallacy at work that can be highlighted at the root of the "impasse" between theology and science related to divine agency in the world. The key logical fallacy is affirming the consequent.[193] In this case the situation is complicated by being developed in a number of stages. The first consequent illogically affirmed according to the gentlemen-clerics and theologians of seventeenth and eighteenth century who studied natural philosophy was:

> If God acts, God acts perfectly.
>
> As God made the world therefore it must be perfect.
>
> Because the world is perfect this proves God acts.

The second consequent affirmed is in similar vein and relates directly to scriptural interpretation and thereby to inspiration. The seventeenth

---

192. Anderson, "Barth and a New Direction for Natural Theology," in Thompson, *Theology Beyond Christendom*, 241–66; Torrance, "The Problem of Natural Theology in the Thought of Karl Barth," 121–35.

193. Warburton, *Thinking from A to Z*, 5–7.

and eighteenth-century students of natural philosophy were convinced they would find perfect order in the natural world as this was the second of God's two books of revelation. The second book could be opened up to human understanding using its own correct methods of study in the same way that the book of Scripture would be opened by correct interpretation. God's action in revelation in creation should be as perfect in its own way as God's action in revelation in the inspiration of the writers of the sacred text was assumed to be perfect. What follows from this is the influence of Darwin and his colleagues who show that the natural world is not a pristine fixed ideal perfection. The conclusion is negated like this: The world is not perfect; therefore, God does not act. Thus, inspiration leading to revelation is not valid, because God does not act. Thus, the whole Christian faith must either be rejected or radically revised.

This train of logic is a fallacy. The negation can only stand if the base assumption is true and the sole possible base assumption, that divine agency must be or can only be by God's perfect action and that such action in humans must be (and possibly in the world) through the *ekstasis* of the soul. Neither the consequent conclusion nor its negation necessarily follows from this particular sequence of steps if an alternative description of divine agency is possible. That is, specifically, if God can act in humans and the world without necessarily being described as reflecting perfect-being theology and that God can act in humans without assuming that they must have an Augustinian metaphysical soul.

It is argued that it is indeed possible to describe divine agency in a way which neither ties it to perfection nor depends on an outmoded metaphysics. The next chapter will return to the proposed incarnational description which begins with considering the pneumatological nature of divine agency within the person of Christ as its basis. This will be examined for theological coherence and plausibility in conversation with the theology of one of the leading figures in twentieth-century Protestant theology, Karl Barth. It is interesting that Barth rejects the Liberal Protestantism to which many had turned in the post Darwinian period. It will therefore be useful to examine his Pneumatology in the conversation with the proposed alternative incarnational description to explore how a Pneumatology might be developed which is not dependent on reflecting the perfections of God's freedom.

CHAPTER 5

Describing Divine Agency in Humans Pneumatologically and Christologically Beginning with Christ

THE CASE WILL BE put that a plausible, coherent and specifically Christian description of divine agency in humans can be developed which is pneumatologically—and christologically—grounded. This description will avoid the problems that Huxley, Darwin and others have identified. These problems, I have argued, developed with describing divine agency in the world as a parallel to that of divine agency in humans. In particular it is that early modern development of divine agency was adapted non-christologically from the Augustinian doctrine of inspiration, early modern perfect-being theology and the two books metaphor. The proposed alternative description of divine agency will not depend on: all things having to have a created purpose; God always having a motive for acting; God always acting for the best; the world being the best of all possible worlds, and; the existence of the soul.

In contrast to the description of divine agency based on Augustinian inspiration, two books and perfect-being theology, the incarnational description as developed to this point makes these distinctive points:

1. divine agency in the world and humans depends on God's choice to act and is not conditioned *a* priori by any property of humanity or feature of the human condition;

2. divine agency in humans is shaped and derives from Christ's continuing reception of the Holy Spirit in his *enhypostatic* humanity without assuming a particular relationship or distinction between the human soul or spirit and the physical;

3. describing divine agency depends solely on theological terminology appropriate to description of the central mystery of the incarnation rather than on terminology first found in philosophical, scientific or medical ideas;

4. applied to inspiration this agency does not automatically require *ekstasis* by the Holy Spirit;

5. the Holy Spirit therefore can act during a broad range of human activities or emotional states in which the Holy Spirit received into Christ's humanity fully preserves the person's humanity as the Holy Spirit acts;

6. the writing of Scripture is not considered as a different class of activity conducted under the inspiration of the Holy Spirit;

7. inspiration by such agency does not automatically guarantee that perfected human action is a result;

Grounding the theological description of divine agency in Pneumatology and Christology in this way may help to resolve the nineteenth-century dichotomy, that the Christian faith must be based on either perfect divine revelation or personal religious affections. This dichotomy, as shown in the previous chapter, developed in relation to the failing fortunes of the assumptions and logic of divine agency in the world drawn non-christologically from early modern understanding of inspiration, the divine perfections and the two books. However, this proposed incarnational divine agency will only make the dichotomy false if it is theologically coherent and plausible while simultaneously resolving or avoiding known difficulties. This coherence and plausibility will be tested substantially through contrasting this proposed description with the Pneumatology of Karl Barth. It will also establish whether it warrants serious consideration as an alternative. If this proposal warrants serious consideration then the dichotomy would be false. The necessity of either the perfection of God's action or the existence of the soul would not predicate God's action in the world. This in turn would open the possibility of renewal of aspects of the dialogue between theology and science where this issue has been a stumbling block.

This engagement with Barth will indicate that some of the claims of the incarnational description can be stated more strongly than above. Some of these claims are more detailed than those assertions made by Barth in his

Pneumatology. In addition, Barth raises additional points not yet considered such as whether the focus of theological anthropology should be about individuals or communities.

Barth's Non-Augustinian Pneumatology

Barth developed his Pneumatology along non-Augustinian lines. The present focus will be on the Pneumatology related to his account of divine agency involving the Holy Spirit. What will be demonstrated is that has not been previously been noted is that Barth remains consistent in his description of the Holy Spirit and the Holy Spirit's work from his earliest lectures on Pneumatology, through the *Dogmatics*, up until *Evangelical Theology*.

Barth's assessment of philosophy and how the natural order should be addressed theologically has important implications for the place of metaphysics, matter/spirit dualism and understanding the relationship of human senses, soul and spirit. The agency by which Christ works by the Holy Spirit in humans is very important to Barth, though he always remains cautious in how detailed theology should be in its description of God, the Holy Spirit. The relationship of the Holy Spirit and humans will be traced in *Romans*, his *Holy Spirit*, and *Church Dogmatics*. This discussion will highlight issues related to the nature and limits on divine action in fallible humans and will directly deal with questions regarding *ekstasis* and error. Barth deals with and reverses the early modern assumptions of the place of inspiration with respect to Scripture and infallibility only after building a case which overturns the non-christological generic Augustinian bases of those early modern assumptions. In particular, Barth's Pneumatology particularly sidesteps the Augustinian assumption of an anatomy assigning functions of reason judgement and direction to the metaphysical soul. It will be shown that Barth's Pneumatology includes the following points that contradict Augustine:

1. There need be no radical dualism of soul and spirit;

2. The agency of the Holy Spirit works in any human actions and is not limited by any human state of consciousness or sanctity;

3. Theology must remain independent of how the world's nature or purpose is understood. Theology must ultimately be independent of anthropological assumptions drawn from the sciences;

4. *Ekstasis* is not automatic. The Holy Spirit does not work by moving the soul of the person to one side but rather by maintaining and enhancing the nature of the person as an individual or part of a community;

5. Differences in degree of inspiration cannot be measured by the results of the action. Barth specifically argues that we have no independent guarantee of the perfection of inspired revelation, rather we must depend on God;

6. The only guarantor of Scripture is God. Scripture reliably achieves whatever purpose that God wants it to achieve;

7. While this does not rule out the infallibility resulting from divine agency in humans, such infallibility or perfection will not meet any independent human definition. Scriptural inspiration does not follow as a special case. Infallibility becomes a non-issue.

These particular features of Barth's Pneumatology cannot be easily found succinctly in his writings needing drawing out. These seven points bear similarity to the contrast already drawn in this chapter with divine agency based in Augustinian inspiration. Engagement with Barth's Pneumatology will do three things. Firstly, some of the claims of the incarnational description will be stronger than they appear at first as Barth shows them to be individual examples of a broader set of requirements of sound theology or Pneumatology. Secondly, Barth cautiously avoids providing the detail of the Holy Spirit's action in humans which the incarnation description might otherwise seem to demand. In providing such descriptive detail, the incarnational description of inspiration will benefit from heeding Barth's reasons for caution. Thirdly, Barth's investigation of related theological questions raises issues that would otherwise be neglected.

Barth attempted to derive a different way past what he saw as the sterile track of naïve Biblicism on one hand and subjective experientialism on the other. He actively questions these trends and their implications in the Protestant theological tradition. Indeed, his thought provides a fertile field for consideration as the question is asked about the implications of freeing discussion of divine agency from the consequences of Augustinian influence and establishing whether such a description may become a workable alternative. Barth is also concerned about finding a place in theology for this engagement with the whole of Christian living.[1]

Initially Barth may seem an odd choice in discussing the dialogue between theology and science. His statement that theology and science have

---

1. Webster, *The Cambridge Companion to Karl Barth*, 6.

nothing to do with each other[2] would seem to end the discussion, echoing Tertullian's, "What has Athens to do with Jerusalem," regarding philosophy. However, Barth has at least as clear an understanding of science and its assumptions and flaws as Tertullian had of the philosophies of his time and their flaws.

Barth's Pneumatology flows throughout his theology including his discussion of science. As we are concerned with the effects of divine agency as one aspect of Pneumatology on the dialogue between theology and science, it is reasonable to expect that Barth's broad Pneumatology could also contribute to theology's dialogue with science. Barth has a distinctly differing opinion to that of both Augustine and Tertullian who claim that *ekstasis* must occur when the Holy Spirit inspires. "Ecstasies and illuminations, inspirations and intuitions, are not necessary. Happy are they who are worthy to receive them! . . . Woe be to us, if we fail to recognise that they are patchwork by-products!"[3] Further by affirming the agency of the Holy Spirit as that which defines our existence, *ekstasis* cannot be an essential element or event or the ultimate beatific vision for Christian living. "The Spirit is the 'Yes' from which proceeds the negative knowledge which men have of themselves. As negation, the Spirit is the frontier and meaning and reality of human life: as affirmation, the Spirit is the new, transfigured reality which lies beyond this frontier."[4]

Barth argues that the work of the Holy Spirit tears down and rebuilds humanity in spirit, understanding and action. The Holy Spirit then works through all the aspects of a humanity thus remade pristine.

> The work of the Holy Spirit, then, does not entail a paralyzing dismissal or absence of the human spirit, mind, knowledge and will. It is often been depicted thus. Attempts have being made to achieve it by strangely resigned twisting of human thought, feeling and effort. It has been overlooked that the attempt to sacrifice the human intellect and will is also an enterprise of the human spirit, that this attempt is impracticable, that the work of the Holy Spirit cannot be forced thereby, and especially that this sacrifice is not well-pleasing to God, that the very intention of the Holy Spirit is to bear witness to our spirit, not to a non-human non-spirit but to the human spirit, that we are the children of God (Rom 8:16), and to help us to our feet thereby.[5]

2. Barth, *Dogmatics in Outline*, 9–11.
3. Barth, *The Epistle to the Romans*, 298.
4. Ibid., 272.
5. CD IV/4:28.

Barth could not more clearly reject the Augustinian description of inspiration or its metaphysical assumptions about human nature. His rejection of these assumptions is of a piece with his rejection of those of Liberal Protestantism as he considers many of these assumptions are shared in common between Liberal Protestantism and science. Barth devotes much of chapters 41 and 42 of *Church Dogmatics* to reappraising the assumptions of Leibniz and Descartes of which divine perfection presents but one of many problems. There are broader issues which intertwine with those related to inspiration, Pneumatology and the dialogue between theology and science in general. This suggests that there is a wider scope of issues beyond just inspiration that have contributed to dissonances between theology and science and within theology.

It is to be noted that Barth rejects the Augustinian presupposition that the image of God must be analogous to some dimension of human beings.[6] This presupposition, Barth claimed, has cast the understanding and description of God in terms of what is generally assumed to be known about human beings. While Barth made this comment about the image of God in relation the doctrine of the Trinity, it also applies to Augustine's doctrine of inspiration and the description of divine agency it contains. Augustine presumes that there must be something which is like the Holy Spirit within the nature of human beings as material body united with a non-corporeal soul and spirit. Augustine's description, as examined in an earlier chapter, assumes an anthropological understanding whose source was not Christian and was capable of being re-expressed without reference to who Christ is. Barth argues that this kind of assumption puts that understanding or philosophy before Christ. That means that knowing God ultimately depends on prior human knowledge—not on divine self-communication.

Barth's comments regarding Augustine are not bluntly dogmatic for the sake of being dogmatic. The key issue is whether human understanding of Christ, the nature of God and the Gospel, should depend on any *a priori*. If they do, then the Christian faith stands or falls depending on whether those assumptions are proved or disproved and, ultimately, not on the person of Jesus Christ. His contention is that first, last and always the Christian faith depends on the person of Jesus Christ, not on a mere philosophy or human theory. If inspiration and divine agency in humans must, as Augustine suggested, necessarily presuppose that human beings are constructed in a particular way, namely with a metaphysical soul that can be stood to one side by God, then inspiration stands or falls on that assumption. However Barth would assert faith depends on the Holy Spirit actually working rather

---

6. Barth, *Göttingen Dogmatics*, 216–18; *"Nein!,"* 80ff: CD III/2:7–13, 80–90.

than on the correct description of that working. Ultimately, if Augustine must be followed, the faith comes to be understood to stand or fall on those assumptions.

These assumptions were shown to fail as the book of nature was read to literally deeper geological layers during the nineteenth century. Also, as scientific knowledge of the function and structure of the nervous system progressed, the need for the soul to be an essentially metaphysical element was challenged. As Huxley highlighted in his work, functions normally attributed to the human soul or spirit had a biological explanation. Therefore, Pneumatology must reject these Augustinian assumptions.

This remains a matter of ongoing contemporary debate in Neurobiology where essentially all functions formerly attributed to the soul as metaphysical can now be explained by biochemical processes—even if they have not actually been proven to operate by these descriptions.[7] Thus, the soul is in danger of becoming the "soul" of the gaps.

Scholarly Debate on Barth's Pneumatology in *Church Dogmatics*

Some studies of Barth's Pneumatology have been attempted. However, none specifically attempt to deal with how Barth describes divine agency in humans. They do however deal with issues relevant to this discussion. The first considered here is that of Rosato, which has been the starting point for debate on Barth's Pneumatology for a number of scholars. Smail, McIntosh, Laats, Pannenberg, Cortez and McCormack have contributed to the discussion. Also, Runia's work is related but raises issues related to Barth's comments on infallibility and the doctrine of Scripture.[8] Runia's work is discussed later.

Briefly, Rosato misses that Barth's Pneumatology is about divine self-communication with the whole person, concluding it is primarily about the mind (noetic).[9] Smail countering Rosato, identifies the integrated nature of Barth's Pneumatology and Christology. Cortez notes the significance of Barth's use of *anhypostasia* and *enhypostasia* in this respect. Laats highlights a significant debate regarding Barth's use of person. He affirms that it is certainly more encompassing than being simply noetic. McCormack notes Barth's Pneumatology as the foundation to his Christology.

---

7. Russell et al., *Neuroscience and the Person*; Worthing, *God, Creation and Contemporary Physics*, 66–71.

8. Runia, *Karl Barth's Doctrine of Holy Scripture*.

9. Rosato, *The Spirit as Lord*.

Why are these issues important? At issue is the coherence and comprehensiveness of Barth's Pneumatology. If Barth's claims made regarding the independence of theology in general cannot be sustained in general it is then unlikely that this could be achieved in the particular case of positing an alternate description of divine agency. Laats' question is whether Barth's theological claims are free of a presupposed anthropology and whether his methodology does in fact make the knowledge of God independent of how humans are understood as a creature or on a preconceived philosophy. These will impinge directly on Barth's understanding of the direct action of the Holy Spirit.

Rosato fails to give a satisfying account of Barth's Pneumatology because he does not appreciate Barth's commitment to careful re-examination of theological assumptions in the light of Christ. Rosato used symbols, with particular pre-existing, self-referential meanings, the antithesis of Barth's method, to explain Barth. In Barth's scheme, the shape of the spiritual nature of humans is of no consequence in coming to know God. One of the more bizarre illustrations that Rosato uses draws from his attempt to cast Barth's approach in the imagery of Teilhard de Chardin.[10] For Barth, Pneumatology must be independent of human constitution, subject only to the Word of God. Humans could be spiritual or not spiritual or even, to be absurd, made in terms of the children's rhyme of "frogs and snails and puppy dog's tails" and this would make no difference to the Holy Spirit's ability to communicate with them. There is also Barth's infamous reference in CD I/1 about the reality of the Holy Spirit's humbling ability to communicate through anything at God's choice. "God may speak to us through Russian communism, a flute concerto, a blossoming shrub, or a dead dog."[11] However, it might be assumed dead dogs do not necessarily come to know God as God.

Barth makes no assumption about humanity upon which theology ultimately would have to be based. The good news of this approach is that with theology independent of what humans might understand themselves to be, no human can be considered to be so confused or so twisted in their understanding as to prevent them coming to a saving knowledge of God.

However, Rosato accurately assesses Barth's conclusion that Neo-Protestantism, Christian Existentialism and Roman Catholicism invariably reduce theology to anthropology despite their good intentions. "Barth's Pneumatology is a reaction to the anthropologising tendencies of

---

10. Ibid., 140.
11. CD I/1:55.

Neo-Protestantism, Christian Existentialism and Roman Catholicism."[12] Reduction to anthropology ultimately makes knowledge of God dependent on anthropology rather than on God. Barth emphatically rejects that knowing God would somehow depend on what humans independently assume they know about themselves. What Rosato fails to appreciate, however, is Barth's stated reluctance in both the Elberfeld lectures and in *Evangelical Theology* to detail the work of the Holy Spirit, fearful of losing a grasp of the person of Holy Spirit in the process describing the detail.

Smail is justified in correcting Rosato, demonstrating Barth's Pneumatology is always linked to his Christology and concerns the whole person.[13] Smail notes that the theology of the Holy Spirit in Barth is something which is woven throughout the text of *Church Dogmatics*.[14] It remains now to tease out these threads, particularly as they relate to the instrument of that agency.

Barth's caution in relation to the Holy Spirit is that the actual self-communication is too easily laid aside for the description. It is true that Barth always links the Holy Spirit to Christ; however, the assertion here is that this is hypostatically linked in a manner more in keeping with the proposed incarnationally constituted divine agency than the modalism that Smail suggests.

Smail skirts around but misses the significance of Barth's use of *anhypostasia* and *enhypostasia*. Barth, however, sees it as impossible to avoid these terms in order to adequately describe the exaltation of humanity in the person of Christ. "But the protest against the concept of *anhypostasis* or *enhypostatis* as such is without substance, since this concept is quite unavoidable at this point if we are properly to describe the mystery."[15] Cortez convincingly summarises Barth's use of *enhypostasis* in criticising descriptions like Smail's.

> Several scholars have argued that the Christomonist objection that Barth subsumes humanity under Christology fails to appreciate Barth's understanding of the "enhypostatic" nature of humanity's relationship to Jesus. Thus, as Jesus' human nature exists enhypostatically in union with the Word so all human nature exists enhypostatically in union with Jesus. Barth's

---

12. Rosato, *The Spirit as Lord*, 21, 53–55, 99.

13. Smail, "The Doctrine of the Holy Spirit," in Thompson, *Theology Beyond Christendom*, 93.

14. Ibid., 87.

15. CD IV/2:50.

christocentricity, then, is more properly understood and the proper ground rather the subsumption of creaturely reality.[16]

Laats also argues that creaturely reality and "person" are grounded in Christology by Barth.[17] Laats makes this comment in summarising a debate questioning whether the nature of Barth's use of "person" was a possible flaw in Barth's theology. It is better stated that Barth leaves the question of personality carefully undefined.

It is only in the later volumes of *Church Dogmatics* that Barth explores the nature of person in the light of God's self-communication. What might appear to be individualistic in CD I is not so by CD IV where the locus of the action of the Holy Spirit is in the Church corporately as Christ shapes humans as community. What Laats correctly identifies is the grounding of the nature of the personal self in the intra divine I-thou relationship between the Father and the Son. As Barth wrote "What we find in the case of the man Jesus is a valid model for the general relationship of man to the will of God."[18] Anthropology cannot be assumed. Therefore, there can also be no assumption of a metaphysical anatomy. The Holy Spirit must be able to work through whatever human beings are in all the limitations of human belief from full understanding to the grossest misunderstanding.

It has been noted earlier that it has often been assumed that Barth's Pneumatology changed from his early to his later writings. However, there are relevant aspects of Barth's Pneumatology that remain constant whilst the focus of his theology grew and developed. McCormack's thesis is that Barth's theology moves from earlier periods marked by dialectic and eschatology to a pneumatological period and then into what he identifies as a thoroughly Christocentric theology which marks the extent of *Church Dogmatics*.

McCormack's scheme puts Barth's work into four phases;

1. First, his initial break with liberal theology as Dialectical theology in the shadow of process eschatology (1915—Jan 1920);

2. The second, dialectical theology in the shadow of a consistent eschatology. (Jan 1920–May 1924)

3. The third, Pneumatocentric Dialectal theology in the shadow of an Anhypostatic-Enhypostatic Christology (May 1924—September 1936)

16. Cortez, "What Does It Mean to Call Karl Barth a 'Christocentric' Theologian?," 141.

17. Laats, *Doctrines of the Trinity in Eastern and Western Theologies: A Study with Special Reference to K. Barth and V. Lossky*, 40.

18. CD II/2:562.

4. The fourth, Christocentric Dialectal theology in the shadow of an Anhypostatic-Enhypostatic Christology (1936-1963) [19]

With reference to the second edition of Romans, McCormack states, "If anything, it was a theology concentrated on the actualization of revelation by the Holy Spirit in the present and therefore was more nearly Pneumatocentric than Christocentric."[20] McCormack notes that focusing on the doctrine of election in Christ as mediator rather than in individual election of the believer, was to make Barth's theology from 1936 onward, "to shift its focus from a Pneumatocentric concentration to a Christocentric concentration."[21] McCormack takes seriously the influence of the Holy Spirit in Barth's work, and corrects a significant trend among twentieth-century theologians to allege Barth as Christomonist who neglected the Holy Spirit altogether.[22] While McCormack does pick up a significant trend in Barth's work, this could be overstated if Barth's christocentric dialectal theology is misunderstood as Barth overlooking Pneumatology. Barth's description of the Holy Spirit remains consistent and is woven throughout his thought. While Barth's focus increasingly moves to Christology, his Pneumatology continues to be a consistent foundation of the remainder of his Christocentric theology. Barth assumes the work of the Holy Spirit as a given and is also consistent in not expounding what he believes should be treated as mystery lest the personal connection in the communication be lost in the description. "A foolish theology presupposes the Spirit as the premise of its own declarations."[23] There is a danger in theologians describing too closely how the Holy Spirit works. The danger being that dependence will be placed on the explanation rather than the person of the Holy Spirit.

Barth was consistently reluctant to describe the inner workings of the Holy Spirit either in the world or in human beings. It is not a simple rejection of clear Pneumatology which would preclude any discussion of a "Barthian" approach to inspiration. It is rather a caution built out of his desire to maintain Christ as the proper subject and object of theology, which is also the work of the Holy Spirit. It also recognises Barth's caution regarding any given system of philosophy. The Holy Spirit acts. The description of the method of action cannot depend on any system of preconceived ideas. For example, Barth was cautious of the millennia long pervasiveness of Augustine's *analogia entis* in western theology. It was the uncritical assumption

19. McCormack, *Karl Barth's Critically Realistic Dialectical Theology*, 21–22.
20. Ibid., 21–22.
21. Ibid., 22.
22. E.g., Grenz and Olson, *20th Century Theology*, 76.
23. Barth, *Evangelical Theology*, 58.

of these Augustinian ideas which he critically assessed in his debate with Brunner in *Nein!*

Natural Theology and Non-Augustinian Pneumatology

Darwin and Barth both see that a problem has arisen in ground of theology. This leads to a limited choice the ground being in biblical infallibility or personal religious conviction. Barth, however, rejects the choice as inadequate. This choice results from problems he identifies in Reformation and pre-Reformation western theology. From an early period in his theology Barth makes a clear break from his theological roots in the nineteenth century's theological academia. Firstly, his analyses provide a useful insight into the assumptions underlying the development of nineteenth-century natural theology, liberal Protestantism and the western academic understanding of the nature of nature. Secondly, Barth's criticism of inspiration may highlight additional problems that must be addressed in order to offer a way around the assumptions which have historically constituted a stumbling block to the dialogue between theology and science.

Barth's departure from nineteenth-century natural theology is linked to his revision of Augustinian assumptions which are related to concepts of divine agency drawn from Augustinian inspiration. This departure became evident in his early work, starting with his well-documented debate with his old teacher and mentor Adolf von Harnack in *Die Chistliche Welt*.[24] It continued in his book on Anselm[25] and in his later debate with Brunner. Barth's correspondence with von Harnack commenced in 1923. His small book on Anselm was completed in 1930 and *Nein!* was written in reply to Brunner in 1935.

Harnack's concern was that Barth, in rejecting scientific historicism, was opening himself and those who followed him to a return to a naïve Biblicism. Harnack's understanding was that the only foil to naïve subjectivism was solid, logical scientific and historical analysis. "Scientific theology is the only possible way of mastering an object through knowledge."[26] In this assumption of a dichotomy between logic and experience Harnack differed little from the approaches to theology shared by many in the late nineteenth century, including Darwin and Huxley. Barth wanted to find an-

---

24. The debate as quoted here appears in full in Robinson, *The Beginnings of Dialectic Theology*, 165–87. It is also well paraphrased with comments by Hunsinger, *Disruptive Grace*, 321–31.

25. Barth, *Anselm: Fides Quaerens Intellectum*.

26. Harnack, quoted in Robinson, *The Beginnings of Dialectic Theology*, 171.

other way forward which re-examined the presuppositions which Harnack had uncritically assumed. What Barth attempted to show is that there is another way, that the contrast between experience and reason is, in fact, a false dichotomy.

Barth sees an inherent arbitrariness in applying any established system of logic or method of investigation. Harnack missed this important point in Barth's argument. What Harnack had uncritically assumed, favours secular scientific intelligibility over theology for as a basis for rationality. Barth does not lightly stand down reason as the final arbiter of faith but argues that final authority in matters of faith cannot be separated from the mysterious action of God in self-communication.

These questions were not merely academic exercises for Barth. He worried that the end consequence of subjective application of reason would lead to a denial of the reality of Christ's command and lead to unethical responses. This would include examples such as Christian theologians justifying war; Harnack being a case in point.[27] Barth's break from his former way of thinking emerged as a result of leading theologians revising ethics in this manner at the outbreak of World War I. Specifically, it was Harnack himself who composed the Kaiser's declaration of war that resulted in Barth's experiencing a personal *Götterdämmerung*.[28] Barth describes reading the declaration of war as, "almost worse than the violation of Belgian neutrality. And to my dismay, among the signatories I discovered the names of almost all my German teachers." [29] Putting reason first as a source for theology would, he realised, lead to an outright denial of the faith when logic, followed to its idiosyncratic conclusions based on limited or flawed knowledge, demands a rejection of God's revealed demands upon all humans.

Countering this, Barth's book on Anselm plays a self-acknowledged seminal role in his theological development. Barth draws from Anselm[30] the idea that rationality should be linked to and derives from the source of theology. Barth adopts for himself this aspect of his summary of Anselm's theological method which colours all his work. That is, no system of logic or rationality can be assumed, without first critically examining it in the light of God's self-revelation in Christ. Two additional foundational motifs that become integral parts of Barth's later work are also highlighted in this work on Anselm. They are, firstly, that the person of Christ is the focus of

27. Barth quoted in ibid., 168.
28. Nebelsick, "Karl Barth's Understanding of Science"; Thompson, *Theology Beyond Christendom*, 175.
29. Barth quoted in Busch, *Karl Barth*, 89.
30. Barth, *Anselm*, 11.

the Christian faith and secondly, that there is a continuous need for the redemption and transformation of human rationality. Barth states that for Anselm the object of faith is identical with Christ. "Not mastering the object but being mastered by it."[31] Barth's method develops into one that actively submits all axioms, preconditions and any well-established conclusions to the critique of the gospel—even to questioning logic and well-established methods. Hunsinger describes the flow of argument as a kind of *aufhebung*[32] which does not merely repeal or negate or annul what has been derived, but reconstitutes them on critical re-examination of any and all conclusions in the light of Christ. His use of Hegelian-like terminology has led to confusion about Barth's method, particularly as much of his work in the early period of *Krisis* theology assumes a Hegelian method of thesis, antithesis and synthesis to arrive at conclusions. In his later work these have to some extent been re-examined and reconstituted in the light of Christ.

Nothing Assumed about the World's Nature

Barth's break with Natural Theology became fully developed in his heated 1935 exchange with his close colleague Emil Brunner.[33] Barth's reply to Brunner was written contemporaneously with writing of CD I/2. What surprised many theologians was a sharp dispute between the two people who initially seemed to constitute the twin representatives of the new neo-orthodox school developing in Germany after World War 1. This debate is significant in the context of this discussion in that Augustine's theological presuppositions are chief among those Barth called into question.

Brunner treats this debate as being about faithfulness to the Reformed tradition. Barth, on the other hand, seeks to correct that tradition taking note of flaws in Reformation assumptions, thus offering a reinterpretation. Brunner's continuing perplexity can be seen in the following comment, "Barth holds the strange doctrine that there is no creature which has in itself any likeness to God."[34] For Barth, it is this analogy precisely that cannot be taken as *a priori*. It cannot be taken as a given or assumed, as the condition for the possibility of undertaking theology. While Brunner asserted that analogy of being was the "basis for every theology,"[35] it would be more precise and in keeping with Barth's reply to describe it as the basis

31. Ibid., 55.
32. Hunsinger, *How to Read Karl Barth*, 98.
33. Barth, "*Nein!*," 65–128; Brunner, "Nature and Grace," 15–64.
34. Barth, "*Nein!*," 53.
35. Ibid., 55.

for every Western Theology. It is Augustine's assumption of the similarity of being between the human spirit and the Holy Spirit which Barth criticises. Specifically, he questions Augustine's assumed similarity of being, which became the basis for necessary *ekstasis* of the human spirit by agency of the Holy Spirit acting in inspiration.[36]

The issue for Barth is not the merits of any particular Reformation doctrine, but rather what *a priori* ideas were being assumed in the Reformers' theology. What Barth seeks to do is to exclude any presumed understanding of the natural world upon which Christian theology must at first be dependent. Barth approaches science and the study of the natural world in a manner consistent with his method so that understanding the natural world is analysed through what is known about Christ. Scientific discovery, artistic intuition and creation, political revolution, moral reorientation and rearmament are not spiritual processes in the sense intended, for, although we may see and understand them as lights which illumine the cosmos as such in reflection of the one light, they cannot be described immediately and directly as self-attestations of Jesus Christ. Even though as secondary lights they are genuine and authentic, and may be recognized as such, they presuppose His self-attestation as the original and true light.[37] This (in Barth) is an effective divorce from the Paleyan programme of determining the divine hand in natural order, which posed such problems for Darwin and Huxley in affirming the Christian faith. The old form of natural theology for Barth is the serpent whose stare hypnotises the theologian until it bites and poisons the very faith he is trying to explain.[38] The difficulty offered by natural theology is that it makes reason, rationality and understanding of history necessarily pre-conditional knowledge for understanding God.[39] God then is subservient to human understanding. How then could the revelation of God lead to the transformation of human thought?[40] The irony is that natural theology is seductive inasmuch as it tried to defend the providence of God, whereas it actually destroyed this providence and sovereignty as the theologian ends up implying that God is the servant of whatever system of reason the natural philosopher presupposes.

Barth offers a distinct break with Augustine. Barth makes the point that the Holy Spirit enacts divine self-consciousness within humans enabling an identity between people and God. Barth states: "We receive the

36. Ibid., 91, 101, 103.
37. CD IV/3.2:501.
38. Ibid., 76.
39. Ibid., 77.
40. Rom 12:2.

Holy Spirit, but our personal identity remains. Who would not agree with that?"[41] The answer to Barth's rhetorical question is Augustine! The action of the Holy Spirit in the person automatically results, according to Augustine, in *ekstasis* of the person's spirit, the seat of identity. In contrast there is, however, no automatic *ekstasis* in Barth's Pneumatology.

Barth maintained that revelation is controlled and sustained by God directly in the person of the Holy Spirit. Barth contradicts Tertullian's argument used in the seventeenth and eighteenth centuries placing priority on the book of nature over that of Scripture. That is, revelation cannot be based on what we know or think we know about the world. Neither, however, Barth argues can God's revelation be found in the gaps of our understanding of the world.

> The miracle of real proclamation does not consist in the fact that the willing and doing of proclaiming man with all its conditioning and in all its problems is set aside that in some way a disappearance takes place and a gap in the reality of nature, and that in some way there steps into this gap naked divine reality scarcely concealed by a mere remaining appearance, of human reality.[42]

In summary therefore, understanding of creation cannot be separated from God's self-communication, Barth argues. The theological understanding of the natural order does not come first but instead depends on God's self-communication, which is, according to Barth, an inherently pneumatological process. Hence, according to Barth, correct understanding of the agency of God's self-communication is vital.

Created Order—Theological Understanding of Creation's Purpose

Existence is, according to Barth, christologically centred and pneumatically sustained. "[T]he self-justification and self-sanctification of God without which He could not have loved the creature nor willed or actualized its existence. . . . the Holy Spirit is the inner divine guarantee of the creature. If its existence were intolerable to God, how could it be loved and willed and made by Him?"[43] Human understanding of existence and the nature of the world is predicated by God's pneumatological self-communicating action,

---

41. Barth, "*Nein!,*" 91.
42. CD I/1:94.
43. CD III/1:59.

which is intimately tied to the question of how this agency is constituted. It is also a pneumatological act which deals with and through human limitations. With this understanding of the theological nature of reality, it is then possible, Barth argues, to deal with the biblical narrative as a human narrative of interaction with God. That is a fully human narrative with all the normal human frailties.[44] That is that they are human accounts used by God to reveal God. What is different is that Barth argues for consistency in God's use of fully human narratives. Where the Augustinian description of inspiration is presumed, the narratives are consequently considered to be perfect beyond human capability.

Barth argues that the notion of created reality is grounded in God's action and not essentially based in human thought or perception of the God revealed in Christ or some generic self-existent perfect-being.[45] The problem for non-christological and non-pneumatological descriptions of the world becomes for Barth not a question of their rejection or acceptance of God but their inability to ask the fundamental "why" question of existence. "All the world-views which have emerged and found some measure of recognition . . . have not yet answered the question prior to all the questions to which they have given their different answers, and that to this extent they have built their systems in the air."[46] Barth believes that this does not exclude theological discourse from active engagement with such systems now or in the future.

> It cannot be the business of theology to decide in what new "dimensions" myth might one day be able to express itself, philosophy to think, or science to investigate; and it is quite improper for theology to assume *a priori* an attitude of scepticism. If a future philosophical system ventures an answer to the prior question, hitherto disregarded or left open or distorted, which underlies all other philosophical questions, then from the standpoint of the Christian doctrine of creation it has not only to be said that this attempt is motivated by genuine necessity, but that it cannot be refused the keenest interest and attention.[47]

Barth sees that there must be conditions for interactions between theological understanding of creation and scientific study of nature. Such dialogue, he argues, is both desirable and necessary. However, a particular understanding of science must not become theology's foundation. While he

44. Ibid., 93.
45. Ibid., 340–41.
46. Ibid., 341.
47. Ibid., 341–42.

argues for the independence of each field of study it would be a mistake to assume he does this because of their independence as disciplines. For Barth, theology is shaped and conditioned by its interaction with Jesus, its subject, through agency of God's self-communication by the Holy Spirit. This is not negotiable. In any interaction with other systems of thought or descriptions of nature, theology must continue to be controlled by this interaction with Christ and remain independent. So according to Barth, Pneumatology and Christology are interwoven in any discussion of cosmology. Barth's approach to the question of world-views has six implications for the Christian doctrine of creation.

1. The Christian doctrine of creation "[C]annot itself become a world-view."[48] It cannot be independent of its source Christ.

2. The Christian doctrine of creation "[C]annot base itself on any world-view."[49] This is consistent with his earlier theological method in that theology must ultimately stand alone on the person of Christ.

3. The Christian doctrine of creation "cannot guarantee any world-view."[50] Its object is Christ and is under no obligation to support any other view.

4. The Christian doctrine of creation "cannot come to terms with these views, adopting an attitude either of partial agreement or partial rejection."[51] Barth argues that while theology supports the goal of providing a comprehensive explanation of the world it cannot be tied to a particular world-view's choice of method and principle.

5. The Christian doctrine of creation in "[i]ts own consideration of these views is carried out in such a way that it presents its own recognition of its own object with its own basis and consistency, not claiming a better but a different type of knowledge which does not exclude the former but is developed in juxtaposition and antithesis to it."[52]

6. The Christian doctrine of creation as a part of Christian dogmatics "pursues its own special task, which is imposed upon it in the service of the Church's proclamation. . . . Understanding the creation of God as benefit it proceeds independently, and is not embarrassed to confess that in

---

48. Ibid., 343.
49. Ibid.
50. Ibid.
51. Ibid.
52. Ibid., 344.

regard to the creation sagas of Genesis, for example, it expects no material and direct help from any world-view, ancient, modern or future."[53]

Barth concludes: "It is for these reasons that the problems posed in natural theology and the philosophy of religion cannot be taken into account in this exposition of the Christian doctrine of creation. Our conversation with the exponents of world-views will be conducted directly."[54] What Barth concludes is that the foundational questions of natural theology are not a starting point for the ongoing discussion between theology and science. In particular, they are not part of the foundation to Christian understanding of the created world. He concludes that it is in science's nature that it must operate out of a presumed set of assumptions. In contrast, Webster described Barth exhibiting "Christian nonconformity"[55] in asserting the separation of theology from science and in particular in asserting theology's independence of any brand of cosmology. Barth wrote: "It cannot fail to strike us that the faith which grasps the Word of God and expresses it in its witness, although it has constantly allied itself with cosmologies, has never yet engendered its own distinctive world-view, but in this respect has always made more or less critical use of alien views."[56] What Webster claims as radical disloyalty to world-views is what Barth has described as the essential discipline of subjecting all theories to critical evaluation in the light of Christ. To fail to do this is to risk theology ceasing to be about faith. Even where "we think we detect an absolute union of faith with this or that world-view, we are not really dealing with faith at all, but with a partial deviation from faith such as is always possible in the life of the Church and of individuals."[57] Theology, Barth concludes, must stand in considered opposition even to those aspects of cosmology whose language must be borrowed in order to make an intelligible statement of the faith in a particular time and place.

> In so far as faith itself is true to itself, i.e., to its object, and in so far as its confession is pure, its association with this or that world-view will always bear the marks of the contradiction between the underlying confession and the principles of the system with which it is conjoined. If there can be no confession of the faith without a cosmological presupposition or consequence (however tacit its acknowledgment), faith can always guard itself against the autonomy of its alien associate. Thus even in

53. Ibid.
54. Ibid.
55. Webster, "Barth, Modernity and Postmodernity," 25.
56. CD III/2:7.
57. Ibid., 9.

these conjunctions of faith with alien world-views its opposition to the latter will always find expression.[58]

It is inevitable that Christian theology will be in association with world-views but must maintain its independence from those world-views, which must remain alien. The process of developing theology without reference to Christology is what is problematic. This discussion of divine agency does not negate Barth's conjecture. It is precisely the baptism of alien metaphysical anthropologies of classical Greek philosophy and medicine that led to problems concerning divine agency in the world in the eighteenth and nineteenth centuries when these were re-expressed in the terminology of generic perfect-being theology. Further, applying Barth's points we may conclude that if theology assumes the notion of the two books of God's revelation we ought not be surprised that this leads to the theologically detrimental persistence of the assumption of human metaphysics and divine perfection in biblical interpretation and theology. For Barth that is, inevitably, the adoption of the two books renders the bible superfluous.

> The sun of the Enlightenment ruthlessly exposed what must always come to light sooner or later when this double system is used. When the two books are juxtaposed as sources of our knowledge of the Creator and creation, it is quite useless to recommend the book of grace. The very fact of this juxtaposition means that the book which is actually read and from which the knowledge of the Creator and creation is actually gained is only the one book, i.e., the book of nature.[59]

If theology is made to stand on these assumptions rather than on God's self-communication by the work of the Holy Spirit the faith is negated. Barth has identified that even laying an *a priori* set of ideas which closely expresses theological truth before God's self-communication eventually separates theology from the truth it seeks to express.[60]

It has been argued here that the understanding of divine agency in the world which developed historically has led to problems as a result of its assumptions about the perfection in the world and human metaphysical anatomy. These have been argued to have arisen from by conjunction of perfect-being theology, the two books and inspiration. The incarnational description of divine agency aims to avoid this specific set of assumptions. According to Barth such absence of assumptions should be a normal part

---

58. Ibid., 10.
59. Ibid., 414.
60. Ibid., 344.

of theological development. It is not that such perfection is in itself undesirable, but rather if the nature of God's control of God's self-revelation is taken seriously then the reasoning behind the logical claim of perfection may not be God's primary concern in God's own self ordering or God's ordering of creation. What if from God's perspective the primary purpose is the act of salvation? That is to say that God's ordering of creation in Christ is firstly to redeem the world or make it redeemable. This would make the Christian understanding of creation ultimately depend on that key christological/pneumatological event, the incarnation.

Holy Spirit and Humanity

It follows from Barth's review of modernity's foundations that problems presented by the Augustinian assumptions inherent in the description of divine agency in the world are not restricted to one particular theological question within theology. Barth maintains that any theological description depending on a world-view is in danger of being rendered invalid if that world-view becomes obsolete. Divine agency as it developed in the eighteenth century depended on the non-christologically expressed notions that the world would reflect the perfections of a generic divine perfect-being and that there existed metaphysical soul. These notions typify an obsolete world-view upon which this theology depends. Thus, such dependence renders this non-christological, generic and metaphysical description incoherent and implausible and implies the same for any other theological description based on this world-view. Here, no less than in any other area of theology, a coherent and plausible Pneumatology needs to be independent of pre-existing world-views.

Barth's Pneumatology was developed in a manner which met these criteria while remaining consistently christological. It is argued that Barth offers a coherent and consistent Pneumatology, which stops short of explicitly linking details of his Pneumatology in a way that allows divine agency to be restated in sufficient detail to remove the assumptions which have been noted as becoming a stumbling block to the dialogue between theology and science. Nevertheless, Barth's Pneumatology is a useful discussion partner for considering whether the proposed incarnational description of inspiration has sufficient scope to warrant serious consideration in the place of the Augustinian description.

Holy Spirit Remaking Humanity as Pristine—Romans

Interestingly, Barth's first major work was primarily pneumatological being about the nature of the Holy Spirit's intersection with humans. The whole content of the epistle of Romans, he writes, "must be understood in relation to the true subject matter which is—the Spirit of Christ."[61] More relevantly to the current discussion, Bultmann saw in Barth's exegesis a rethinking of the form of the dogma of inspiration. Barth unashamedly noted this—in that he assumes that the Holy Spirit will somehow speak through the written words.[62] The notion that God must somehow and in some way self-communicate is a key idea Barth expressed here in 1921 as well as later at Göttingen (1922-1923) and Elberfeld (1929). Writing a commentary was for Barth a process that aimed to bring the reader to hear and be negated and renewed by the reality of God's self-communication.

Intersection of the Divine and Human—Krisis

Krisis was Barth's earliest description of this relationship in God's self-communication, was of the divine "yes" meeting our human "no." "The relation between us and God, between this world and His world, presses for recognition, but the line of intersection is not self-evident. The point on the line of intersection at which the relation becomes observable and observed is Jesus."[63] The exact nature of this intersection and its relationship to humanity in general Barth did not detail in *Romans*. The Holy Spirit touches the world tangentially in the resurrection, touches without touching.[64] In like manner human beings are touched by the presence of Christ. The possibility of contact between God and humanity is possible only through the work of the Holy Spirit who bridges the inconceivable differences in the person of Christ. God impinging in *Krisis* changes our understanding of the world without however changing God.[65] How interaction between the Spirit and the human occurs remains a mystery appreciated in paradox.

The sharp distinction between human and divine which requires the *Krisis* of God is highlighted for Barth in his discussion of Abraham in the assertion that no one has been a "directly visible divine-humanity or

---

61. Barth, *Romans*, 17; Barth, *Die Rommerbrief*; Barth, *The Göttingen Dogmatics*; Barth, *The Holy Spirit and the Christian Life*.
62. Barth, *Romans*, 18.
63. Ibid., 29.
64. Ibid., 30.
65. Ibid., 82.

a human-divinity."⁶⁶ The demonstration of righteousness in humans only comes as a result of Christ illuminating the human. Every time that Barth uses the term *Krisis* throughout *Romans*, he implied that the paradox welds together incomparable opposites, joined by the continuing agency of the Holy Spirit. Throughout his commentary the ability of the human to obey the command of God comes only through the *Krisis* action of the Holy Spirit. This is a constant element of his Pneumatology from his earliest work.

Revelation comes by the continuous action of the Holy Spirit within the human person.⁶⁷ Barth links this intimately with the death and resurrection of Christ.⁶⁸ Only the Holy Spirit can enable the truth of the death and resurrection of Christ to be known to an individual as "Truth." Outside of faith this truth is not known. Only in the inseparable action of Spirit and Christ can the reality of salvation be known to humans. Barth describes it as the person being simultaneously nullified and preserved, (*aufgehoben*) by God.⁶⁹ This is a re-examination and preservation process which is not limited to assumptions or knowledge in Barth's theology but is also the basis for the transformation of human beings so that everything about a human is both brought to nothing and reconstituted whole in the presence of Christ. This is the Holy Spirit's work in the human person.

Humanity Nullified and Preserved in the Spirit

It is interesting that Hoskyns chose to translate the terms *aufgehoben, aufheben* and *aufhebung* as "dissolve" and "dissolution" in the English translation of *Romans*. The translation of these terms solely in this way is potentially misleading. What is missed is Barth's point that it is "dissolution" in preparation for reconstitution on a higher plane."⁷⁰ The dictionary definitions of *Aufgehoben, Aufheben* and *Aufhebung* are accepted as, "preserve" or "be in good hands," "abolish" or "annul," and "abolition" or "annulment."⁷¹ Their use appears Hegelian. Nevertheless, the image of the Holy Spirit "dissolving" our humanity in the light of Christ in *Krisis* does convey a strong if somewhat incomplete image in the English. The action of the Holy Spirit in Barth's usage does not end there but continues on making all assumptions

66. Ibid., 118–19.
67. Ibid., 72.
68. Ibid., 159.
69. Ibid., 95, 116; Barth, *Die Rommerbrief*, 69, 91.
70. Barth, *Romans.*, xiv; Hunsinger, *Disruptive Grace*, 98.
71. Sasse, Horne, and Dixon, "Cassell's New Compact German—English English—German Dictionary."

about our humanity, life, relationship to God totally new. It remakes as pristine what it is to be human.

Commenting on Paul's injunction to be renewed in mind,[72] Barth makes further reference to the centrality of *aufhebung* in the action of the Spirit. "It is precisely the THINKING of the thought of eternity which dissolves (*aufhebung*) the possibility of any adequate human thought."[73] This critical revision of the preserved human reason by relationship with its primal origin in God which is enacted by agency of the Holy Spirit is a point which Barth returns to repeatedly.[74]

While the place of the mind is important in the work of the Holy Spirit, according to Barth, this work encompasses the whole of human existence. In this approach, human existence remains ambiguous as it depends on the personality, even the whim of God. Conversely, this "whim" of God is a source of hope which has its source in the faithfulness of God. Humans need to live in freedom as Barth enjoins: "Oppressed on all sides by God and wholly dissolved (*Gott aufgehoben*) by Him; reminded constantly of death and as constantly directed towards life."[75] "Wholly dissolved" should also read God preserved. These actions of the Holy Spirit control and determine the transformative self-communication of God in the Holy Spirit, which he later supposes will encompass the whole of human life and action.[76] This requires some exposition lest it be concluded that at this point Barth has eliminated all possibility of describing the immediate action of the Holy Spirit. Rather it is that ultimately humans cannot control such a description. Humans may describe how God works, but that description must always be controlled by the one who is described. Theology becomes a science in Barth's usage where its theories are controlled by the object of its study.

Barth's treatment of the action and relationship of the Holy Spirit with humanity echoing Calvin's *mysterium* is not a simple avoiding of the question. In his eyes, it is the recognition of our inability to hold the revelation of God as our own apart from the intimacy of direct contact with God. He expresses it thus: "When we try to find the content of divine Spirit in the (pardoned) consciousness of man, are we not like the man who wanted to scoop out in a sieve the reflection of the beautiful silvery moon from a pond? What can or shall we find there to investigate?"[77] The action of the Holy

---

72. Rom 12:2.
73. Barth, *Romans*, 437–38; Barth, *Die Rommerbrief*, 422–23.
74. Barth, *Romans*, 476, 482–83, 503.
75. Ibid., 503; Barth, *Die Rommerbrief*, 487.
76. CD IV/2:125; CD IV/1:21.
77. CD I/1:216.

Spirit is that which opens the human to the truth of God. "In the work of the Holy Spirit this man ceases to be a man who is closed and blind and deaf and uncomprehending in relation to this disclosure affected for him too. He becomes—a man who is open, seeing, hearing, and comprehending."[78] Consistently in *Church Dogmatics,* the knowledge of God, of revelation, is linked in some way to the divine/human union of the incarnated Christ. This link is what I have suggested as constitutive in the proposed incarnational divine agency.

Barth concludes that God is knowable, that the action of the Holy Spirit is perceptible, but rightly understood it is God who controls what humans know. This is occurs through and controlled by the inspired witness of Scripture. This is a witness humans are unable to leave if they are to know God. What distinguishes Barth's theology (at this point in Barth's development) from post eighteenth century conservative theologies is that his approach to scriptural authority is not expressed in terms of the reliability or perfection ensured by inspiration but rather by God's direct and continuing action. Barth's startling claim is that this agency of the Holy Spirit revealing Christ brings with it its own epistemology which becomes the foundation for human understanding and action, communal, individual and ethical. Barth refers to ethical obedience to God as "the new divinely inspired reality in the midst of other realities."[79] Thus, Barth actively treats inspiration in the daily living of Christian life as a continuation of the same act of God in the inspiration of Scripture. While in no way describing the substance of spirit, Barth affirms the nullification and preservation of human self-knowledge and existence as the work of the Holy Spirit. Inspiration is thus for Barth an ongoing part of the Holy Spirit's divine agency—a continuous work of revealing God in the person.

What is highlighted is the risk entailed in having too good a description of the Holy Spirit, of depending on the explanation rather than the person of God. "Our fear of denying the Spirit is far greater than our fear of betaking ourselves to the ambiguous and questionable realm of religion."[80] Assumptions make the interaction of spirit and human seem an impossible paradox. These dualities of paradox are negated and renewed by the Spirit. The mystery of the agency of the Holy Spirit is that what is inevitably paradoxical to human understanding is not to God. It is only in the presence of God that the paradox is resolved by negation and renewal. This is the

---

78. CD IV/4:27.
79. Barth, *Romans,* 260.
80. Ibid., 274.

"God-given riddle" which is the mystery of which humans should not be ignorant.[81]

Elberfeld—Holy Spirit Bearing Humanity

It is often assumed that Barth's Pneumatology changed with his change in terminology from his early work to his later work. However, there are aspects of Barth's Pneumatology relevant to this discussion of divine agency that remain constant. While Barth discontinues use of the term "Krisis," the agency of the Holy Spirit as negating and renewing humanity remains a constant theme upon which his theology rests. This are clearly identifiable in his 1929 *Holy Spirit*.

In the lectures given at Elberfeld, Barth described: The Holy Spirit as Creator; The Holy Spirit as Reconciler; and the Holy Spirit as Redeemer. The first question is, "How can God be known?" Barth states that human beings are not in a position from which they can survey their relationship with God and say that they are open upwardly toward God. Therefore God's self-communication by the Holy Spirit is to be considered only as given to the creature from,

> Him-who-exists-for-us, and so it is not a *datum* but a *dandum* (Lat: "to be given"), not as fulfilment but as promise. Grace is our having been created, but it is also "created for God." But grace is ever and in all relations God's *deed* and *act*, taking place in this and that moment of time in which God wills to be gracious to us, and is gracious, and makes his grace manifest. It is never at all a quality of ours, inborn in us, such as would enable us to know of it in advance. Every other interpretation signifies, covertly or overtly, in the premises of conclusions, another interpretation of the Holy Spirit than as the creative power of our own spirit.[82]

That is, apart from God's gracious continuing action any other interpretation of the Holy Spirit, either covertly or overtly, signifies that interpretation depends on what is in some ways already known. The givenness of revelation is a theme repeated constantly in the first lecture. It is the work of the "Holy Spirit to be continually opening our ears to enable humans to receive the Creator's word."[83] The Holy Spirit turns our internal and external

---

81. Ibid., 413.
82. Ibid., 5–6.
83. Ibid., 8.

response to commands in Scripture into God's command for people. The creator's word turns

> "urges" into a command of God. These two things—the presumed sure knowledge about the divine compulsions of our own existence and the confident taking of the Bible, as if it gave a list of moral counsels—are both in principle identically arbitrary. ... [T]hose outer and inner constraints of our existence must be ever *acquiring* the character of divine indications, duties and promises through the divine speech to us.[84]

He concludes that ethics is pneumatological in that it can be neither a description of commands nor an appeal to truths supposed to lie in nature or divine text. The Holy Spirit works by continually inspiring divine self-communication in people. Revelation is a gift, in which the hand of the giver must never leave the gift. If it is taken as a gift and becomes the possession of the receiver, then that person, rather than becoming a visionary, becomes the fanatic. Barth provocatively suggests the danger becomes more real though less apparent the better or the "more Christian," the received and owned revelation becomes.

> He might, of course, always hear something, and he always supposes that he hears something in what has been said by God to him—something that he can make a start with, call it, if you like, a view of the universe according to his particular penchant, whether conservative or revolutionary. On the basis of this he fancies he has inside information about himself and can thus direct and control both his own and others' lives. And just then—and the neater, the more confirmed, the more practical, the "Christian" this program of his turns out well—all the more surely has he missed the Word of the Creator.[85]

The danger is that the more explicit the description of the Holy Spirit's action becomes, then the more likely it is that these propositions replace relationship.

The Holy Spirit is not some ideal spirit of the good, the beautiful or the true, but the "incomprehensible" Spirit who acts against human hostility to God to bring about real fellowship between God and sinful humanity. It is the Holy Spirit who overcomes sin in the human, and the overcoming of sin must be of a piece with how the agency of the Holy Spirit within the human person. Barth suggests that any attempt to overcome sin without the action

---

84. Ibid., 9.
85. Ibid., 10.

of God is in itself sinful as it is done without God. Thus, "When man's own action, whatever its pretence or form, is made into a condition with regard to fellowship with God, then the Holy Spirit has been forgotten, then sin will be done to overcome sin."[86] Barth sees this as a Pelagian linking of an assumed human ability and as inseparable from Augustinian theologies. "The Protestant doctrine of grace in the new age is almost entirely a variety of the Augustinian theme . . . [T]his means 'It is divine gift and man's creative action combined in one.'"[87] Barth's harsh rejection arises from dismissing the notion that the action of the Holy Spirit leaves some imparted quality that becomes the possession of the human. This is the same idea as in the first lecture regarding the impossibility of imparted revelation which might become the possession of a false visionary. For Barth reconciliation, repentance and righteousness without relationship removes their reality.

Barth continues to make the action of the Holy Spirit in sanctification a similar process to that in the Holy Spirit's work of God's self-communication. Similarly, Barth grounds the human/divine nature of the Holy Spirit's action not in a simple analogy with the combination of human and divine in Christ, but points back to the reality of the combination of sinfulness and justification grounded in the redeemed humanity of the person and action of the Word of God.

> Indeed, the Christian is *simul peccator et justus*, and the surmounting of this irreconcilable contradiction does not lie in the Christian—not even in the most secret sanctum of his existence, nor does it happen in any of the hours of his life's journey, not even in those hours most moved and profound, of the conversion and death—but it is the action of the Word of God.[88]

Barth does not describe divine agency in the world or in humans as being constituted in the person of Christ as has been proposed in this book. Nevertheless, what Barth propounds is in keeping with the proposed incarnational description's implications. The life of faith is not one that inheres in the human. Thus, there is no hypostatic union of the sinful and the divine in the Christian which gives raise to the life of faith. "Faith, together with its experience of judgement and of justification, is God's work: totally hidden and pure miracle. But on this account it is no *hypostasis*, hovering over or in front of or behind the actual man."[89] These things derive and are defined,

---

86. Ibid., 20.
87. Ibid., 22, citing *Enchirdon* 32.
88. Ibid., 30–31.
89. Ibid., 32.

argues Barth, from and by God. It is the Holy Spirit who continually constitutes human life as Christian life.

In his third lecture Barth moves on to discuss the role of the Holy Spirit in the purpose and end of humanity. He deals with Augustine's views because they are

> the classical representation of the Catholic view, which in Protestantism even, either avowedly or secretly holds the field in both Catholic and Protestant theology. This is the view of man as one existing in presupposed continuity with God. This view of continuity between God and man is always threatening to make man out as being his own creator and atoner.[90]

His concern is to avoid any notion that humans can in some way become their own savior or creator in the here and now of the present. Barth sought to preserve both a description of God as both full sovereign and relating to humanity in full condescension (as stated in his opening remarks of these lectures). The purpose and end of humanity becomes swept up in the love and life of God in which every aspect of life, action and hope is defined and enabled by God.

> This life lived through the Holy Spirit becomes a life lived in hope, like Abraham's. Above and within the tangled tokens of our creaturehood in "the kingdom of nature," above and within the warfare of the spirit with the flesh in the "kingdom of grace," there is an ultimate, an Immovable and a Final End in the "kingdom of glory." Is there *it*?—nay: is there a *He*? Yes: God the Holy Spirit. And this Ultimate, this Actuality, which he gives, is, is ever-coming, never already come; always manna for today, never to be kept for tomorrow and the day after: it is the Christliness of the Christian life.[91]

In his Elberfeld lectures, Barth makes explicit the essential link between Pneumatology and ethics implied throughout *Romans*. The Holy Spirit is the one whose consistent action enables revelation, reconciliation, repentance and righteousness to become manifest in an otherwise hopelessly sinful and ignorant humanity. The human is enabled by the Holy Spirit's agency to live a life in intimate relationship with God expressed in devotion, obedience and prayer.

> The wonder of prayer—and this is a thing quite different from the "infused grace" of ability to pray aright—is the incoming

90. Ibid., 60.
91. Ibid., 64.

of the Holy Spirit to the help of the man who is praying. It is the Spirit's sighing, which to be sure, is in *our* mouth; yet as *his* groaning, who creates out of the man who is sober or drunken or finical, or even the *homo religious* who has utterly collapsed (I mean by that, the man who prays in himself and to himself); out of a man of that kind, the Holy Spirit makes a person who *actually, really* prays.[92]

The ultimate expression of the Holy Spirit's action is a life lived in prayer, because "it had pleased God to take this groaning, sighing man, together with his burden, upon *himself.* This grave circumstance is the presence of the Spirit of Promise."[93] Barth's description of Pneumatology does not vary over his career. His reluctance to describe how it is that the Holy Spirit acts within humans also does not vary. While the incarnational description is not incompatible with Barth's description of the Holy Spirit's work, it goes into more detail than Barth by claiming that divine agency in humans is constituted by the pneumatological/christological action of the Holy Spirit in Christ. Barth was reluctant to explore the Holy Spirit's interaction with the human spirit more closely lest its mystery and the relationship be lost. This is a limitation which constitutes a good reason for caution in the current discussion.

Holy Spirit Dealing with Unbelief, Heresy, and Presuppositions

Barth notes that in its very nature, theology must deal with conflicting opinions. "If it rests on a conflict in which faith finds itself, then, if this conflict is to be serious, it must be a conflict of faith with itself."[94] Theology must firstly be able to deal with unbelief as this is what prompts theological expression as a corrective. He claims that resolution of such conflict cannot adopt "particular forms in which it finally becomes only too clear to the opposing partner that is either deceiving him when it proposes to deal with him on the ground of common presuppositions, or that it is not quite sure of its own cause in so doing."[95] Theology, he claims must begin with its own critical re-evaluation. All of its presuppositions must be modified so that the whole might be preserved or renewed. This means an active engagement

92. Ibid., 68.
93. Ibid.
94. CD I/1:31.
95. Ibid.

with those theologies which have become too one-sided in relation to particular theological presuppositions, i.e. heresy.

> In true encounter with heresy faith is plunged into conflict with itself, because, so long and so far as it is not free of heresy, so long and so far as heresy affects it, so long and so far as it must accept responsibility in relation to it, it cannot allow even the voice of unbelief which it thinks it hears in heresy to cause it to treat it as not at least also faith but unbelief.[96]

Theology itself must be able to seriously deal with human fragmented, contrary and mistaken understanding, being itself a human response prompted by the work of the Holy Spirit to describe the self-communication of God. The controversial stand Barth takes is that even the great post-Reformation trends in theological orthodoxy are not in themselves free of heretical tendencies, because they in turn have not seriously taken account of their own presuppositions. "We stand before the fact of pietistic and rationalistic Modernism and rooted in mediaeval mysticism and the humanistic Renaissance. The fact of the modern denial of revelation, etc., is quite irrelevant compared with this twofold fact.... [T]he Evangelical faith stands in conflict with itself."[97] Not only must the Holy Spirit's inspiring work be unlimited by conscious state or sanctity, this work must be unlimited by humans who are in error or far from God. While the Holy Spirit may provide varying degrees of improved or even perfected human action, the Holy Spirit's ability to act is not limited by human performance or lack thereof.

If there are no guarantees of perfection, the question then is how humans can be sure they know anything about God and their true relationship with the world. There can be, according to Barth, no philosophical pre-conditioning of theological rationality. Theology is first of all subject to the will of God which expressed to particular people in specific times and places. Scripture

> is there and tells us what is the past revelation of God that we have to recollect. It does so in the first instance simply by the fact that it is the Canon.... Gk:kanon means rod, then ruler, standard, model, assigned district. In the ecclesiastical vocabulary of the first three hundred years it was used for that which stands fast as normative, i.e., apostolic, in the Church, the *regula fidei*, i.e., the norm of faith, or the Church's doctrine of faith."[98]

96. Ibid., 33.
97. Ibid., 34.
98. Ibid., 101.

God's interaction is remembered and there is a standard set of recollections to measure future interaction by—the canon.

Barth's Doctrine of Scripture

Barth begins his doctrine of Scripture pointing to the normative nature of specific historical revelation, whatever might be made of it in the present. Barth does not begin with inspiration. Instead, his Pneumatology more broadly informs his doctrine of Scripture. As Scripture is normative, how it is to be used and interpreted and applied in the life of the Church is important. Barth warns of the dangers of trying to control revelation by the method of interpretation:

> The Church goes astray in respect of the Bible by thinking that in one way or the other it can and should control correct exposition, and thereby set up a norm over the norm, and thereby captures the true norm for itself. The exegesis of the Bible should rather be left open on all sides, not for the sake of free thought, as Liberalism would demand, but for the sake of a free Bible.[99]

In addition, he warns of the danger of making the text alone the controller of revelation without the action of the Holy Spirit working and applying it:

> We thus do the Bible poor and unwelcome honour if we equate it directly with this other, with revelation itself. This may happen when we seek and think we find revelation in the heroic religious personality of the biblical witness. But it may also happen in the form of a general, uniform and permanent inspiredness of the Bible.[100]

In his *Göttingen Dogmatics*, Barth had dealt with the place of the seventeenth-century description of verbal inspiration or the "direct dictation" description of inspiration. The problem he states is that this description makes revelation become direct revelation. The bible becomes a "paper pope."[101] It makes revelation a gift possessed independently of its giver, so that the one who possesses the gift comes to depend on the gift rather than the giver who is God. By Barth's reasoning, the self-communication of God witnessed to and controlled by the Scriptures must reveal the very person

---

99. Ibid., 106.
100. Ibid., 112.
101. GD 1:217.

of God. It is not enough to reduce revelation to mere concepts or facts that might have some independent existence on their own—that is, having a being apart from or beyond God. The location for making a final judgement about what is revelation must be, he argues, removed from the sticky grasping hands of humanity. This would set up "a real and effective barrier" against treating Scripture as "a fixed sum of revealed propositions."[102] The trouble with human attempts to grasp revelation is that they invariably impose preconceived notions upon the reality of God's self-communication and filter that self-communication. Consequently, this makes what is received less than who God is by reducing revelation to ideas, philosophies or feelings.

While not rejecting reason and logic, even these too Barth believes must be examined to their foundations. "Even our knowledge of the Word of God is not through a reason that has somehow remained pure and that can thus pierce the mystery of God in creaturely reality."[103] Barth moves his description of revelation beyond being tied to mere intellectual facts or elicited religious experience. "It is wholly through our fallen reason. The place where God's Word is revealed is objectively and subjectively the cosmos in which sin reigns. The form of God's Word, then, is in fact the form of the cosmos which stands in contradiction to God."[104] Thus, he moves beyond the dichotomy that tested Darwin and Huxley. It is inevitable, however, that people will hear revelation where they are and as who they are and in their culture with imperfect logic and reason. Barth is confident that God chooses to speak even through their imperfection or even in spite of their imperfections.[105]

To trust that God is acting, revealing, and working by the Holy Spirit is an act of faith. The good news which is really good news, as Barth sees it, is that God graciously chooses to connect with people as they are. Given that humans need to acknowledge the control of God in revelation while wrongly but unavoidably continuing to separate understanding of revelation from God's self-communication, how did Barth describe a doctrine of Scripture and its role in Christian life?

Barth's doctrine of Scripture attempts a withdrawal from both liberal historico-critical use and conservative Biblicist use of the text. The suspicion that Barth levels at any foundational philosophy also extends to assumed historiography and presumed exegetical hermeneutics. The method used

102. CD I/1:7.
103. Ibid., 166.
104. Ibid.
105. Ibid., 202.

to interpret Scripture is less important to him than the fact or event that God uses and works through that interpretation. "Thus the judgment that a biblical story is to be regarded either as a whole or in part as saga or legend does not have to be an attack on the substance of the biblical witness."[106] He attempted to distance himself exegetically from the presuppositions of both approaches, while attempting description of how the Holy Spirit communicates God's self through the text to individuals. He had a very low view of the development of these schools. In a critique of their weaknesses as philosophies, he states: "Because this is not the case, the philosophy of religion of the Enlightenment from Lessing by way of Kant and Herder to Fichte and Hegel, with its intolerable distinction between the eternal content and the historical 'vehicle,' can only be described as the nadir of the modern misunderstanding of the Bible."[107] The text confronts us and is in God's hands and is out of human control. Barth's doctrine of Scripture sees the inspiring role of the Holy Spirit as central to the way that God is disclosed. This action of God is for him a bald fact, independent of any understanding of anthropology. Such schemas were potential minefields of assumptions whose alienness has been unrealised.[108]

Humans are faced, in Barth's judgement, with the impossibility of trying to contain the actuality of God if they try to hold revelation in their own minds. Attempting this also denies the personal aspect of this divine self-communication in the Holy Spirit. Only God can hold this mystery together in the human being.

> And now we repeat that the God of the biblical revelation can also do what is ascribed to Him in this respect by the biblical witnesses. His revelation does not mean in the slightest a loss of His mystery. He assumes a form, yet not in such a way that any form will compass Him. Even as He gives Himself He remains free to give Himself afresh or to refuse Himself.[109]

Only God can make the incomprehensible comprehensible. Because revelation and, implicitly in Barth's scheme, the agency of the Holy Spirit in human beings are entirely controlled by the action of Christ: anthropology or anatomy cannot be presumed. For Barth, theology must be independent of what humans believe they know about themselves independently of God's self-communication. Therefore describing the action and working of the Holy Spirit and the role of the Holy Spirit in revelation according to

---

106. Ibid., 327.
107. Ibid., 329.
108. Ibid., 343.
109. Ibid., 324.

Barth must be possible while remaining independent of Augustinian anthropology, the metaphysical soul and automatic *ekstasis*.

Just as the knowability of God cannot be based in Barth's description on any presumed human anatomy, similarly the guarantee of revelation does not rest on any assumed perfection of the book of the Word of God. This freedom of God to act in relation to humans is also, according to Barth, expressed in how the Holy Spirit acts in the human to free the person for life lived with God. The conjunction of the Holy Spirit and the person in this freedom enables the person to become a real recipient of revelation, the self-communication of God. "The problem by which we found ourselves confronted was: How can man believe? How does *homo Peccator* become *capax verbi divini* (sinful man become capable of the divine word)? The New Testament answer is that it is the Holy Spirit who sets man free for this and for the ministry in which he is put therewith."[110] The "how is it possible?" question, is the theological question of divine agency. What is the agency by which the Holy Spirit actually works in some way in or through people? This question is at its heart about how divine and human actually meet and are expressed. Barth's answer to his question lies in the Holy Spirit bringing to life within the sinful person the resurrection life of Christ. It is a transition from death to life. In this, the action of the Holy Spirit follows the action of the resurrection.

> This is what it means to have the Holy Spirit. To have the Holy Spirit is to be set with Christ in that transition from death to life. The form of a real recipient of God's revelation, the form which gives the law to his thought, will and speech will always be the form of the death of Christ.[111]

The Holy Spirit has become by the action of Christ "at home" in the sinful person not in some "higher" spiritual or metaphysical element of a person.

> But this also means that the creature to whom the Holy Spirit is imparted in revelation by no means loses its nature and kind as a creature so as to become itself, as it were, the Holy Spirit. Even in receiving the Holy Ghost, man remains man, the sinner. Similarly in the outpouring of the Holy Ghost God remains God.[112]

Because of the Holy Spirit's work change in the nature of the person does not come to belong to the being of the person. There is no requirement

---

110. Ibid., 456.
111. Ibid., 458.
112. Ibid., 462.

for *ekstasis* to happen as the Holy Spirit works. There is, however, a requirement that this agency be continuous. Barth leaves foundation of the agency of the Holy Spirit through people a mystery. Barth terminates his description of how the Holy Spirit works within the human person at this point.

Barth instead turns to the mystery of Christ. That is it is a mystery not because of simply being mysterious or hidden but rather because the comprehension of God is fundamentally beyond the capacity of human finitude.

> In the first place, it refers back the mystery of the human existence of Jesus Christ to the mystery of God Himself, as is disclosed in revelation—the mystery that God Himself as the Spirit acts among His creatures as His own Mediator, that God Himself creates a possibility, a power, a capacity, and assigns it to man, where otherwise there would be sheer impossibility. And the mention of the Holy Ghost is significant here in the second place, because it points back, to the connexion which exists between our reconciliation and the existence of the Reconciler, to the primary realisation of the work of the Holy Spirit.[113]

This preservation of humanity in the Godhead, whilst at the same time preserving the dignity of the Godhead, is the essence of *anhypostasis* and *enhypostasis*. The mutual preserving of the two natures in Christ becomes the act upon which Barth bases his doctrine of Scripture.

> In recent times the doctrine of the *anhypostasis* and *enhypostasis* of Christ's human nature has occasionally been combated by the primitive argument, that if the human nature of Christ is without personality of its own, it is all up with the true humanity of Christ and the Docetism of early Christology holds the field.[114]
>
> The Miracle of the incarnation, of the *unio hypostatica*, is seen from this angle when we realise that the Word of God descended from the freedom, majesty and glory of His divinity, that without becoming unlike Himself He assumed His likeness to us, and that now He is to be sought and found of us here, namely, in His human being.[115]

Barth argues that in the person of Christ, Christ himself takes on sinful human flesh and redeems it. This point at "issue is simply this, whether Christ's flesh had the grace of sinlessness and incorruption from its own

---

113. CD I/2:199–200.
114. Ibid., 164; parallel CD IV/2:49.
115. CD I/2,:165; parallel CD IV/2:64–65.

nature or from the indwelling of the Holy Ghost; I say the latter."[116] In keeping with the second part of Anselm's *Cur Deus Homo*, Barth argues that Christ incarnate finally makes true God's word about what humanity was created to be—in himself. Christ in the incarnation remakes as pristine what it is to be human. This happens in the person of Christ by the action of the Holy Spirit. It is an act of "pure grace." Rather than explaining the incarnation and consequently the agency of the Holy Spirit in terms of being supernatural miracles presupposing an understanding of natural law, Barth argues, "The mystery does not rest upon the miracle. The miracle rests upon the mystery. The miracle bears witness to the mystery, and the mystery is attested by the miracle."[117] It is the action of the Holy Spirit which makes the revelation of God real to us in God's self-communication of God's very self.

It is this self-communication grounded in the incarnation which is the basis for Barth's doctrine of Scripture. Even though Barth has argued for God's preservation and use of human frailties and even the sinfulness of the writers of Scripture,[118] he still argues for the highest view of Scripture. "In this respect also, we must realise the adequacy of Holy Scripture as the source of knowledge. We must submit to our bondage to Scripture. We must submit to be content with it. We must do so no less because man is in the very presence of God."[119] God is present through the text when God chooses to be there, to work by and act through the Holy Spirit in the writer and in its hearers. He notes that freedom to hear God is God-given and not "freedom from man's side."[120] Barth's doctrine of Scripture makes indivisible the linkage between Scripture, revelation, inspiration and the direct agency of the Holy Spirit. Just as these are indivisible he further argues for a similar indivisible link between the Church and the Holy Spirit. Not that God is bound by some form of hypostatic union with the Church but that God chooses inexorably to self-communicate via this means. "Barth locates God's self-communication specifically in the Church.[121] While revelation is by God's choice, it is inexorably linked to the "warts and all" reality of the Church. "While God is as little bound to the Church as to the Synagogue, the recipients of His revelation are."[122] It is in turn indivisibly linked with human dealing with the text, but it remains ultimately in the control of God

---

116. CD I/2:154.
117. Ibid., 202.
118. Ibid., 353.
119. Ibid., 208.
120. Ibid., 209.
121. Ibid., 362.
122. Ibid., 211.

and is only realized in confession and at the direct action of the Holy Spirit. That is that the production of sacred text, as well as its editing, preservation, retelling, translation, interpretation and application requires a continuation of divine action—inspiration by the Holy Spirit.

While Barth does not make a direct constitutive connection between this pneumatic Christology and the agency of the Holy Spirit in humans, to suggest that the latter derives from the former is a key aspect of the incarnational description of divine agency suggested earlier. To conclude that the divine/human nature of the Holy Spirit's inspiring work derives from the divine/human nature of Christ extends Barth's description without endangering the mystery. What appear to be similar mysteries in the agency of the Holy Spirit's actions can be described as resulting from being constituted by the central mystery of Christ's incarnation. The action of the Holy Spirit in the incarnation of Christ is the derivative link to the agency of the Holy Spirit in people.

Barth on *Theopneustia*

Barth having dealt with his doctrine of Scripture then turns the biblical question of *theopneustia*,[123] from which the term inspiration derives.[124] Barth examines 2 Timothy 3:14-17 and 2 Peter 1:19-21 closely. In relation to the first passage Barth states: "The whole Scripture is literally: 'of the Spirit of God,' i.e., given and filled and ruled by the Spirit of God, and actively out breathing and spreading abroad and making known the Spirit of God. It is clear that this statement is decisive for the whole."[125] In relation to 2 Peter 1, Barth adds the following:

> The "prophecy of Scripture" is rightly read in the sense of what precedes; it is our light in a dark place, when it is act made the object . . . : i.e., when we allow it to expound itself, or when we allow it to control and determine our exposition. This is because, as the text goes on, it is not given "by, the will of man," but in it men spoke as they were "moved by the Holy Ghost," they spoke "from God."[126]

What is interesting about Barth's linking of the two passages is that he ties the breathing of the Holy Spirit to God's choice to reveal God's very self

---

123. Lit., God-breathed.
124. See the earlier discussion in 2.1.
125. CD I/2:504.
126. Ibid., 504-5.

in a specific time and place. It is his assertion that the 2 Peter reading is the clearer of the two texts which confirms the force Barth gives his argument. The *theopneustia* of Scripture is linked decisively to the ongoing prophetic work of the Holy Spirit in the Church. "The decisive centre to which the two passages point is in both instances indicated by a reference to the Holy Spirit, and indeed in such a way that He is described as the real author of what is stated or written in Scripture."[127] According to Barth, God's choice to respect the authors' humanity does not diminish their role as authors. "As though in what they spoke or wrote they did not make full use of their human capacities throughout the whole range of what is contained in this idea and concept."[128] Their action was their own but, "was surrounded and controlled and impelled by the Holy Spirit, and became an attitude of obedience in virtue of its direct relationship to divine revelation—that was their *theopneustia*."[129] Thus according Barth, the biblical concept of *theopneustia* points "to the event which occurs for us: Scripture has this priority, it is the Word of God. But it only points us to it. It is not a substitute for it. It does not create it. How can it, seeing it is only a description of what God does in the humanity of His witnesses?"[130]

The authority the God-breathed text has is the authority which God gives it. It does not have authority deriving from assumptions requiring perfect divine action in its writing. "The men whom we hear as witnesses speak as fallible, erring men like ourselves. What they say and what we read as their word, can of itself lay claim to be the Word of God, but never sustain that claim."[131] The irony is that in the west since the Reformation the Protestant Church has in its creeds both affirmed and gone beyond Augustine. In doing this, it has overlooked the essential point:

> We know what we say when we call the Bible the Word of God only when we recognise its human imperfection in face of its divine perfection, and its divine perfection in spite of its human imperfection. In relation to the obvious uncertainty of the traditional Canon, whether in respect of its compass or its textual form, this could be conceded by many writers.[132]

The trouble, Barth argues, has been that in Protestant theology the tendency has been to focus on one aspect of Pneumatology rather than

127. Ibid., 505.
128. Ibid.
129. Ibid.
130. Ibid., 505–6.
131. Ibid., 507.
132. Ibid., 508.

the whole in discussing inspiration.[133] The one-sidedness of focusing on the produced text leaves particular problems as it limits discussion for the whole of God's self-communication and reduces the text to a datum humans can grasp. This narrow focus leaves an inability to address significant Early Church thought which ascribed inspiration to non-canonical and even pagan texts. "What are we to think of Theophilus of Antioch and Pseudo-Justin when they did actually ascribe the same inspiredness to the prophets and the Sibyllines?"[134] The difficulty with post-Reformation Protestant orthodoxy is the inability to address the breadth of what the early Church considered inspired. To speak of inspiration only as verbal inspiration is according to Barth an extreme cheapening of the richness of the biblical and patristic concepts of inspiration. To reduce inspiration to mere words and to focus solely on the inerrancy of grammar, or concepts, and to restrict this to the historically remote production of the texts reduces the value of what God continues to do by the Holy Spirit to virtually nothing.

Infallibility and Docetism

The historical tendency to reduce discussion of inspiration to mere transmission of ideas, and the consequent focusing solely on infallibility of the pristine text has other complications. These run afoul of the problem of becoming human-controlled which Barth had already criticised. These are, he contends, attempts to reduce the incarnation to something graspable and totally understandable in human thinking.

> Where there is this idea of a "dictation" of Holy Scripture through Christ or the Holy Spirit, is not the doctrine of inspiration slipping into Docetism? If I am right, it was Augustine who first spoke clearly about a divine dictation. . . . But if it was not intended docetically, how else can we think of it except again on the Jewish and heathen model—as a mantically-mechanical operation? And if it is not to be regarded as mantically-mechanical, how can it not be docetic? The same choice is even more stringently imposed when we are told by the second-century Athenagoras that the Holy Spirit moved the mouths of the prophets as his organs. . . . But the price which had to be paid for this apparent gain was far too high. By, as it were, damping down the

---

133. Ibid., 517.
134. Ibid.

word of man as such, by transmuting it into a word of man . . . the whole mystery was lost.[135]

Barth sees this reduction of mystery as signifying that "already the doctrines of inspiration of the Early Church were leading to a rather, naive secularisation of the whole conception of revelation."[136] Barth highlighted what he saw as problems resulting from the Reformation period handling of the Augustinian inspiration. The understanding of inspiration was now transformed,

> from a statement about the free grace of God into a statement about the nature of the Bible as exposed to human inquiry brought under human control. The Bible as the Word of God surreptitiously became a part of natural knowledge of God, i.e., of that knowledge of God which man can without the free grace of God, by his own power, and with direct insight and assurance.[137]

Barth argues the eighteenth and nineteenth-century developments tried to correct the perceived naivety of the necessarily miraculous perfection of the textual originals. In his view, however, this did not change anything. Barth is radical in suggesting that the change in doctrinal formulation is merely a change from one form of attempting to control the biblical text to another. The end result is the same, by controlling it, real revelation is effectively denied.

> "This doctrine of inspiration was absolutely, new." But it was so, not in its content, which was merely a development and systematisation of statements which had been heard in the Church since the first centuries, but in the intention which underlay the development and systematisation. As we have seen, the earlier statements were not free from ambiguity. They did not escape the danger of a docetic dissolving or of a mantico-mechanical materialising of the concept of the biblical witness to revelation.[138]

Barth argues these developments obscure revelation and ultimately obscure true knowledge of God. In addition to having had to deal with the dictation description of inspiration this also brought Barth to focus on assumptions about perfect divine action.

135. Ibid., 518.
136. Ibid., 519.
137. Ibid., 522–23.
138. Ibid., 525.

Infallibility and Fallibility

Barth argued that if the authority of Scripture is to be derived from being perfectly dictated, it follows that the defence of Scripture:

> Had to be stated with this almost terrifying pedantry and safeguarded against all possible defections, we always come up against the postulate that Holy Scripture must be for us a *divina et infallibilis historia*. Truth is necessarily diffused over all Scripture and all parts of Scripture ... Should there be found even the minutest error in the Bible, then it is no longer wholly the Word of God, and the inviolability of its authority is destroyed.[139]

There is nothing new in Barth's logic—if inspiration means perfect dictation, scriptural authority is all-or-nothing. It must be miraculous. The positing of Scripture as miraculous puts argument about the nature of Scripture in a strange place in which the truth stands or falls because of an alleged fact which is not demanded by the text, whereas, "genuine, fallible human word is at this centre the Word of God: not in virtue of its own superiority, of its replacement by a Word of God veiled as the word of man, still less of any kind of miraculous transformation, but, of course, in virtue of the privilege that here and now it is taken and used by God Himself."[140] God takes and uses human words. The argument for infallibility implies the potential of shame in the presence of error which God does not share. "If God was not ashamed of the fallibility of all the human words of the Bible, of their historical and scientific inaccuracies, their theological contradictions, the uncertainty of their tradition, and, above all, their Judaism, but adopted and made use of these expressions in all their fallibility, we do not need to be ashamed."[141] God no more demands of human language that it be perfect before working through it, than demanding that people be perfect before they can be saved. "Verbal inspiration does not mean the infallibility of the biblical word in its linguistic, historical and theological character as a human word. It means that the fallible and faulty human word is as such used by God and has to be received and heard in spite of its human fallibility."[142]

The question of infallibility is a red herring for Barth in terms of how he described the authority of Scripture. Scriptural authority is a given, which is established by God's choice. "Consequently the Church cannot evade Scripture. It cannot try to appeal past it directly to God, to Christ

---

139. Ibid., 524.
140. Ibid., 530.
141. Ibid., 531.
142. Ibid., 533.

or to the Holy Spirit. It cannot assess and adjudge Scripture from a view of revelation gained apart from Scripture and not related to it."[143] Barth later returns to the issue of infallibility in CD IV/1 and again picks up what he sees as a negation of revelation's God-givenness in favor of human authority.

> We must not forget that the transition from biblical to biblicist thought does involve the transition to a rationalism—supernaturalistic though it is in content. Therefore the relationship of theology to the truths of revelation which it has taken from the Bible is no longer the relationship to an authority which is superior to man. It has fundamentally the same assurance and control with regard to them as man as a rational creature has in regard to himself ... believing that he is the master of himself.[144]

Runia, a conservative Reformed theologian, is confronted by Barth's doctrine of Scripture.[145] Ultimately, Runia's argument is often sterile as he misses some essential elements of Barth's method while defending the traditional Reformed doctrine of Scripture, in the form of the notions of infallibility and biblical authority. Runia misses Barth's re-examination of assumptions and rebuilding of the whole of the theological programme. Barth's defence of Scripture as a fully human document implies fallibility which confronts Runia who defends Scripture's infallibility as a matter of first importance.[146] However, the fallibility versus infallibility debate is a secondary issue for Barth as it arises out of Augustinian assumptions which he dismisses at the outset. The notion of infallibility arises as a logical conclusion of Augustine's doctrine of inspiration. Specifically Runia needs the witness of Scripture to be reliable to ensure Scripture's reliability and authority. What he misses in Barth is that the witness of the author or of the receiving community of faith is not the guarantor that Scripture is the Word of God. Only God is.

The traditional justification for infallibility has depended on the Augustinian description of inspiration, which makes an assumed necessarily perfect divine action the basis for the text's lack of error. This justification is undermined by the Augustinian description assuming an understanding of human anatomy which is no longer tenable. Flawed Augustinian inspiration can no longer be the guarantor of infallibility. The problem, which faces those theologians who choose to defend infallibility, is that they now cannot argue for infallibility on the basis of *ekstasis* inspiration. Also, if infallibility

143. Ibid., 544.
144. CD IV/1:368.
145. Runia, *Karl Barth's Doctrine of Holy Scripture*, 1–168.
146. Ibid., 58, 77.

could or should be retained it must be defended by a more circuitous path which makes no assumption about human metaphysics and works in spite of human limitations and error.

Runia does pick up and agrees with the emphasis that Barth places on keeping Pneumatology and Christology together, but does not see Barth's key as the paired notions of *anhypostasia* and *enhypostasia*. Ironically, he agrees with Barth that all Docetism and monophysitism must be entirely removed from the doctrine of Scripture, but disagrees with suggestions from his contemporaries that the traditional Reformed doctrine of Scripture is Monophysite.[147]

Barth has no doubt that the doctrine of the verbal inspiration

> was not merely worked out as a bulwark against a growing rationalism, but that it was itself, not an expression of an over-developed faith of revelation, but a product of typical rationalistic thinking—the attempt to replace faith and indirect knowledge by direct knowledge, to assure oneself of revelation in such a way that it was divorced from the living Word of the living God as attested in Scripture.[148]

This authority is according to Barth, something which never leaves the hand of God even though in real terms in the Church people seek to clarify and debate the issues involved. It is in God's anticipation of human need and reaction and in God's choice to act directly in the full scope of human activity that "Scripture demonstrates its freedom anand supremacy in the fact that, above and beyond the power of resistance and criticism, it has the power of assimilating and making serviceable to itself the alien elements it encounters."[149]

Scripture rightly understood comes to humanity as a gift out of the mystery of the incarnation which demonstrates how God uses it, the shape of divine grace to a sinful people. Thus it is possible to conclude that Barth argues that the only guarantor of Scripture is God. Scripture reliably achieves what God wants it to achieve. Scriptural inspiration does not and must not be treated as a special case. While Barth does not rule out the infallibility of God's action during inspiration, such infallibility or perfection is not automatic and will not meet any assumed human definition.

---

147. Ibid., 72n48.
148. CD IV/1:368.
149. CD I/2:682.

Barth's Anthropology and Holy Spirit

Barth on the Traditional Distinction of Soul and Body

Rather than being a special case of the action of the Holy Spirit, the development and transmission of Scripture must be, according to Barth, an ongoing action of the Holy Spirit in the usual manner of God's self-revelation. The preservation and perfection of the human nature by God in the revelation and self-giving of Christ begs the question of what constitutes human nature in relation to the spiritual makeup of a human. "Through the Spirit of God, man is the subject, form and life of a substantial organism, the soul of his body—wholly and simultaneously both, in ineffaceable difference, inseparable unity, and indestructible order."[150]

Explicating the traditional body/spirit duality, Barth saw a problem, which more recent understandings exacerbated.

> Advance into the region of new insights and conceptions which we now enter has its own difficulties. For one thing, we come at this point very close to the propositions of all kinds of non-theological studies of mankind, among which one can very easily go astray, especially as they always arouse at this point the burning interest which powerful inner contradictions always bring to light.[151]

Barth, discontent with such understandings, attempted to recast the traditional duality of a human as body and spirit in union from a theological perspective. What Barth understands to be the traditional western view of duality of spirit/soul and body reduces to these four key points:

1. "That man has Spirit means that God is there for him. Every moment that he may breathe and live he has in this very fact a witness that God turns to him in His free grace as Creator."[152]

2. "That man has Spirit is the fundamental determination which decisively makes possible his being as soul of his body."[153]

3. "Since man has Him, the Spirit is certainly in man—in his soul and through his soul in his body too. It is the nearest, most intimate and

---

150. CD III/2:325.
151. Ibid.
152. Ibid., 362–63.
153. Ibid., 363.

> most indispensable factor for an understanding of his being and existence. But while He is in man, He is not identical with him."[154]

4. "The Spirit stands in a special and direct relationship to the soul or soulful element of human reality, but in only an indirect relationship to the body. The soul therefore is the life of the body, and therefore the human life as such which man may not only have but be when he receives the Spirit."[155]

One important rider to this description is that God as Spirit remains independent of the human and that the human does not become God. While this is Barth's assessment of the tradition view, he was not happy with it.

Body and Soul—Barth Rejects the Classical Duality

Barth is content with neither regurgitation nor his careful revision as he rejects this duality. "We necessarily contradict the abstractly dualistic conception which so far we have summarily called Greek, but which unfortunately must also be described as the traditional Christian view. According to this view, soul and body are indeed connected, even essentially and necessarily united, but only as two 'parts' of human nature."[156] The trouble, Barth points out, with this traditional description even when refined is that it does what he accuses more recent liberal theology of doing—blurring the lines between "science" and theology.

> Our argument against it is simply that its conception does not enable us to see real man. Man is also, and indeed wholly and utterly, body. This is what we must be told by materialism if we have not learned it elsewhere. But there is no sense in trying to seek and find man only in his body and its functions. For if he is really seen as body, he is seen also as soul, that is, as the subject which gives life to his material body.[157]

It is the whole person that is in every part real, whereas the categories of Greek philosophy in effect make some virtual. One difficulty is the ease by which this type of conception married itself to early Christian theology. An alien world-view in effect became the foundation for further theological

---

154. Ibid., 364.
155. Ibid., 365.
156. Ibid., 380.
157. Ibid., 383.

development. Then in circular argument, theology came to support this baptized world-view.[158]

While arguing against body/soul dualism Barth indicates that a simple commitment to materialism is insufficient as it fails to appreciate a comprehensive nature of humanity. "Reaction against the Greek traditional Christian dualism can, however, come from quite another quarter than materialism. Indeed, monistic materialism past and present obviously calls for the counter-attack of a monistic spiritualism, which takes the opposite view that the soul is the one and only substance of human reality."[159] The dualism between body and soul and the resolution of this into a monism of either extreme are all problems as Barth sees it. "The abstract dualism of the Greek and traditional Christian doctrine, and the equally abstract materialist and spiritualist monism, are from this standpoint a thoroughgoing and interconnected deviation"[160] This gives rise to the view of the human as "a totality composed of two parts inadequately glued together, of two obviously different and conflicting substances."[161] Only in God does the complex nature of humanity make sense, Barth claims.

The knowing self cannot be considered a spiritual abstraction of a given dualism between the soul and body. Self-consciousness and perception must be holistic and not divided between the corporeal body and metaphysical soul. Barth excludes the categories upon which Augustine's anthropology and ultimately his notion of inspiration depend. Barth details his rejection of such notions asking the question where does perception begin and end. "Superficially we can recognise the two moments of human nature, body and soul, in the division of the idea of perception into awareness and thought."[162] The Augustinian categories cannot apply:

> There can thus be no question of a simple distribution of the two functions in the act of perception to soul and body, or of the simple notion of co-operation between the two. The situation is rather that man as soul of his body, is empowered for awareness, and as soul of his body for thought. Understood thus, the two are different and cannot be interchanged.[163]

---

158. For a contemporary discussion see, Green, "Restoring the Human Person," in Russell et al., *Neuroscience and the Person: Scientific Perspectives on Divine Action*, 3–22.
159. CD III/2:390.
160. Ibid., 393.
161. Ibid.
162. Ibid., 400.
163. Ibid.

Similarly, Barth argues for the material participation of the body in thought as well as perception. "Again, thinking is not only with the soul. How could his soul think, if it were not the soul of his brain, his nerves and his whole organism? Even when he thinks, man lives the life of his body."[164] While Barth refers to Jesus as the one in whom humanity is perfected, he notes, "The biblical view of man which has been our guide and which we must now consider in detail. We remember that we shall search the Old and New Testaments in vain for a true anthropology and therefore for a theory of the relation between soul and body." In this Barth continues to be echoed in recent dialogue between theologians and neuroscientists. The interdependence of cognitive functions formerly thought to be separated between soul and body is highlighted in a series of papers presented at a joint Berkley and Vatican conference in Poland in 1997. Many functions such as emotions feeling and habit which were in the past attributed to the soul have been identified as located in biochemical and neural structures.[165]

Barth's affirms the silence of Scripture on a theory for the relations between soul and body, rather there an essentially holistic unity of the human. Barth's theological "agnosticism" regarding the soul parallels Huxley's questions about the soul, its existence and location. However, where Huxley leant toward denying its existence as traditionally described, Barth argues for its redefinition without assuming a metaphysical dualism. Therefore, when a human being has the Spirit it is to be understood that, "this means that the view of the active unity of soul and body is here fundamentally secured."[166] If this "fundamentally secured" the nature of humanity as an essential unity by God and with God is understood, this implies the agency by which God chooses to interact with the human. "If man understands himself in his relation to God as established and ordained by God, in relation to soul and body as the two moments of his being he can in no case understand himself as a dual but only as a single subject, as soul identical with his body and as body identical with his soul."[167]

With this understanding, the agency by which the Holy Spirit enters the situation is not as a foreigner, but as sustainer and perfector of human reality. Any description of the agency of Holy Spirit in humans which automatically requires *ekstasis* would require the Holy Spirit to always negate

---

164. Ibid., 401; Russell et al., *Neuroscience and the Person;* LeDoux, "Emotions: How I've Looked for Them in the Brain," 41–44; Arbib, "Towards a Neuroscience of the Person," 77–100; LeDoux, "Emotions—A View Through the Brain," 101–18; Jeannerod, "Are there Limits to the Naturalization of Mental States?," 119–27.

165. CD III/2:433.

166. Ibid., 434.

167. Ibid., 426.

what God chooses to sustain as its essential reality. Barth however foreshadows an additional element in the discussion of how the Holy Spirit acts in humans. That is, the Holy Spirit acts in humanity. Just as the individual is not a human divorced from the sustaining presence of God, neither does the individual exist divorced from the redeemed human community in that same sustaining presence. "The community lives under the lordship of Jesus in the form of the Spirit. In the Spirit that double proximity is actual presence."[168]

Barth argues not only that there is be no duality of soul and spirit for individuals but that God acts on humans as a whole and that God actually redefines the understanding of anthropology so that there is no duality of soul and spirit in humanity. Therefore, theological discussion about the nature of humans cannot be separated from the nature of the whole of humanity. The action of the Holy Spirit shapes and sustains the very existence of humanity.

Humans as Community

But what is humanity, is it fundamentally a collection of individuals or a community? The role of the Holy Spirit as Barth describes it is not limited to the individual but becomes part of God's sustaining of the community of humanity and in particular the redeemed community within humanity—the Church. It is "the Spirit who is given not as the Spirit of the individual but as the Spirit of the whole community."[169] When this is considered, the emphasis in the present context must be upon the fact that it is this community which is called into being by the Holy Spirit.[170] Christ in controlling the shaping of the Christian as part of the self-communication of revelation is extended by Barth to the broader community.

> From all this it is self-evident that neither the Christian community nor the individual Christian can subjugate or possess or control Him, directing and overruling His work. He makes man free, but He Himself remains free in relation to him: the Spirit of the Lord. He awakens man to faith, but it is still necessary—to believe in Him, in *Spiritum Sanctum*.[171]

168. Ibid.
169. CD III/3:255.
170. CD IV/1:654.
171. Ibid., 646-47.

This extension from the individual nature to the communal nature of the Holy Spirit's interaction with humanity is not an incidental aspect of Barth's Pneumatology.

> In modern times, under the influence of Pietism, we have come to think in terms of the edification of individual Christians—in the sense of their inward inspiration and strengthening and encouragement and assurance. The cognate idea has also arisen of that which is specifically edifying. Now all this is not denied. It is, indeed, included in a serious theological concept of upbuilding. ... The New Testament speaks always of the upbuilding of the community. I can edify myself only as I edify the community.[172]

Barth argues in Pneumatology that there need be no radical dualism of soul and spirit and that it must be independent of any and all anthropological assumptions. Nonetheless, Barth having described the Holy Spirit's working in community makes a judgement against individual pietism. By denying or at least limiting the scope individual divine action, Barth exhibits a problematic turn of logic. Barth seems to suggest that humanity is to be understood as communal rather than individual. Surely, the theological description of the action of the Holy Spirit needs to be independent of whether the basic unit of humanity is the individual or community? Otherwise knowing God depends not on God but on how humanity is understood. Maybe in reframing Pneumatology not only should there be no assumption of body soul dualism but also no assumption of a duality between individuality and community.

Extending discussion of the agency of the Holy Spirit into the realm of community is an important addition to Pneumatology by Barth. It also provides an additional reason why Barth does not explore more precisely the question of how the Holy Spirit interacts with people.

Why Barth Stops—Mystery and Holy Spirit

Whilst examining Barth's Pneumatology it has become apparent that he applied his method to the systematic re-evaluation of its theological, scientific and philosophical presuppositions. Barth also desired to describe Pneumatology as a consistent whole. Both of these objectives were noted as necessary to offering a revised description of divine agency. Up to the place in Pneumatology beyond which Barth will not pass, the proposed incarnational description of divine agency in humans does not clash with Barth's

---

172. CD IV/2:627.

approach. Barth does stop developing detail in his description of how the Holy Spirit works in human persons. Barth's reason for not going further, signals an important caution for theological discussion. The question here is whether Barth develops his Pneumatology in a way that might provide some assistance to supporting the incarnational description of divine agency.

Within the severe limits Barth sets it is possible to elicit what his description of divine agency looks like. Noting these limits and the shape of his description of divine agency it may be asked whether the proposed alternative incarnational explanation has merit. After summarising what may be learnt from Barth's Pneumatology and in what shape it may be considered unfinished, it remains to show how moving beyond where Barth stops is necessary in order to enable removal of divine agency as a stumbling block in the dialogue between theology and science.

Barth's Limit

While Barth extended the action of the Holy Spirit to the human community rather than merely individuals, Barth saw that no description of the action of the Holy Spirit could be adequate. "In this context there can be no question of an exposition of Acts 2 which has regard to all the dimensions and problems of the passage. We shall consider only from the standpoint which is our present concern. And for that reason we shall only touch the fringe of its true mystery."[173] Barth describes the Holy Spirit's action as a mystery, but in this mystery the action establishes the reality of the community. He sees the pattern as decisive that God's ongoing actions should be appreciated in mystery rather than comprehensively understood. "Self-evidently this speech and its success are meant to be regarded as the work of the Holy Spirit as He is imparted to the disciples and constitutes them the community."[174]

The Holy Spirit acts decisively and powerfully. However, Barth claims that the lack of explanation about how the Holy Spirit acts is deliberate. Referring to Peter's speech and the gift of tongues Barth continues, stating that the Holy Spirit "makes men capable of this authoritative and effective witness, and therefore the true mystery of Pentecost, is only revealed at this point. Yet it is the obvious intention of the narrator first to introduce this mystery in outline and without explanation, differentiating it from all similar mysteries which might occur to the reader."[175] The claim is that such an

173. CD III/4:320.
174. Ibid., 321.
175. Ibid.

action can only be understood in outline but not in detail. "Luke's account of this miracle was indispensable, not to explain this miracle, which speaks for itself, nor to enhance or establish its historicity, but to limit and define it."[176] It is the lack of detail in the biblical accounts that Barth uses to support his inference that investigation and description should stop. What some strands of theology seek to describe with precision; Barth argues can only go so far and no further. Acknowledging that the action of the Holy Spirit is most important in knowing God, Barth asks the question why is it necessary to go further and why investigation should stop.

> There can be no higher or deeper basis of knowledge, or revelation, than the witness of the Holy Spirit, who is the Spirit of Jesus Christ the Lord. Why then can we not rest content with this bare reference to Him? We obviously cannot do this if we have rightly understood and explained it as a reference to the Holy Spirit who as the Spirit of Jesus Christ renders witness to Him, His self-witness.[177]

The first issue for Barth is that of simple abstraction. If the witness of the Holy Spirit is listened to as a witness' account then, he argues, people will indirectly try to approach God. "If we listen to the witness of the Holy Spirit and give it its proper place, we find that we are not referred directly, but very indirectly, to the One who attests Himself in it. Indirectly! But this means that those who accept the witness of the Holy Spirit cannot tarry with Him as such. There can be no abstract receiving and possessing of the Holy Spirit."[178] Secondly, the Holy Spirit points only to Christ and so provides no self-disclosure.

> The witness of the Holy Spirit does not have itself either as its origin or goal. It has no content of its own. It has no autonomous power. It does not shine or illuminate in virtue of its own light. The Holy Spirit may be known, and distinguished from other spirits, by the fact that He does not bear witness to Himself.... It is the fulfilment of His self-witness.[179]

The mediation of the Holy Spirit which allows divine action in humanity is a mystery that Barth argues derives from the mediation of the Spirit within the life of God.

---

176. Ibid., 322.
177. CD IV/2:130.
178. Ibid.
179. Ibid.

> The answer which we now make is that it is because in this mystery of His (Holy Spirit's) being and work in our earthly history there is repeated and represented and expressed what God is in Himself. In His being and work as the mediator between Jesus and other men, in His creating and establishing and maintaining of fellowship between Him and us, God Himself is active and revealed among us men.... It takes place first in God Himself. It is an event in His essence and being and life.[180]

Barth stops because he believed theology dare go no further. Though, as will be shown, he notes dissatisfaction with this. This mystery of the Holy Spirit's mediation, which Barth seeks to protect, cannot in human understanding be separated from that of the incarnation. The Holy Spirit's mediation can only be experienced humanly as a result of the incarnation. The incarnational description of the agency of the Holy Spirit posed earlier argues this very derivation is constituted by the pneumatological/christological preservation of the two natures of Christ. In proposing this, the incarnational description extends in detail beyond where Barth stops, but rather than removing the mystery, relocates it to Christ.

Barth's Understanding of Divine Agency

Where Barth stops in Pneumatology limits the detail which can be ascribed to his understanding of divine agency in humans. Nonetheless, it is God's direct action which can never be understood apart from the fact that it happens. It is dependent on God and assumes, indeed needs to know beforehand, nothing about anthropology. Within these limitations that Barth sets for himself and the whole of Christian theology, he does make a statement describing divine agency in answer to the following question. "There remains only the question of the manner of His working and therefore of the development of the power and lordship of Jesus. How does the Holy Spirit act? How does He encounter us? How does He touch and move us? What does it mean to 'receive' the Spirit, to 'have' the Spirit, to 'be' and to 'walk' in the Spirit?"[181] He notes the temptation to avoid an answer, but also notes the inadequacy of any answer.

> But how does this avoidance, or obscure description, harmonise with the fact that in the Holy Spirit, although we do have to do with God, we do not have to do with Him in His direct being in

180. Ibid., 341.
181. Ibid., 360.

> Himself, which might well reduce us to silence or allow us only to stutter and stammer, but with God (directly) in the form of the power and lordship of the man Jesus?[182]

Barth's description of the agency of the Holy Spirit comes couched in a warning against overplaying even this description. The Holy Spirit acts from Christ "man to man."[183] "But in the relationship between the man Jesus and other men, in the exercise of His power and lordship, and therefore in the operation of the Holy Spirit, this is not one possibility among many, nor is it merely the norm, but it is the only reality."[184] It is independent of feeling or peculiar action at the time. "To receive and have the Holy Spirit has nothing whatever to do with an obscure and romanticised being."[185] The agency of the Holy Spirit ethically moves humans toward obedience, "To be or to walk in Him is to be under direction, and to stand or walk as determined by it. . . . [T]he work of the Holy Spirit is always distinguished by the fact that it is and gives direction: the concrete direction which proceeds from the man Jesus."[186]

While Barth affirms that the agency of the Holy Spirit comes through the humanity of Christ, he says nothing about christological anthropology in relation to this agency. Thus he does not affirm the nature of Christ's reception of the Holy Spirit into his own humanity as constituting how the gift of the Holy Spirit is communicated to fallen humanity. The action of the Holy Spirit is, however, that which perfects and opens the human to the truth of God.

> To put it again in a single sentence: In the work of the Holy Spirit the history manifested to all men in the resurrection of Jesus Christ is manifest and present to a specific man as his own salvation history. In the work of the Holy Spirit this man ceases to be a man who is closed and blind and deaf and uncomprehending in relation to this disclosure affected for him too. He becomes— a man who is open, seeing, hearing, and comprehending. Its disclosure to all, and consequently to him too, becomes his own opening up to it. In the work of the Holy Spirit it comes about that the man, who with the same organs could once say No thereto, again with the same organs, in so far as they can be used

---

182. Ibid., 360–61.
183. Ibid., 361.
184. Ibid., 361–62.
185. Ibid., 362.
186. Ibid.

for this purpose, may and can and must say yes. In the work of the Holy Spirit that which was truth for all, hence for him too.[187]

While Barth's pneumatological account of divine agency stops short of the incarnational description proposed earlier, his explanation picks up many of its essential elements in terms of source (Christ) and outcome on the whole human being.

Barth Unfinished?

How complete was Barth's Pneumatology? Barth argued that theological formulation must be independent of any anthropology presuming individuality or community. Barth's movement from individualism is evident in the vocation of the Church in CD IV/2. However, this is not fully explored. Near the end of his life, Barth indicated that even though *Church Dogmatics* would not see completion, nevertheless he had essentially covered all that he had wanted to address in the bulk of his other writing.[188] The next logical step would be applying the method of *aufhebung* to self and community in Christ by the action of the Holy Spirit. This would have been a consistent Barthian logical next step for his opening discourse for the never written CD V. Note has already been made that his Pneumatology in *Evangelical Theology* at the end of his career is similar in wording and structure of argument to his early work. It is entirely possible that Barth saw no need for major revision in his Pneumatology and its intimate relationship with Christian ethics. Rather what it lacked was thorough exposition. Elsewhere, he viewed his Pneumatology as a "theology which now I can only envisage from afar, as Moses once looked on the Promised Land."[189] He knew what the shape and the promise of the land would be but was unable to explore it fully.

What is unfinished—and deliberately so—is a detailed description of the agency of Holy Spirit in humanity. Barth affirmed the internal testimony of the Spirit but saw serious dangers in efforts to be more precise about inspiration than "God speaks to us through the Bible."[190] In *Evangelical Theology*, Barth states that explaining how inspiration works carries the danger of thinking "we know whom has the Spirit" rather than depending on the

---

187. CD IV/4:27.
188. Barth, *How I Changed My Mind*, 65–68.
189. Busch, *Karl Barth*, 494.
190. Kraeling, *The Old Testament since the Reformation*, 169; Barth, *Evangelical Theology*, 54.

Spirit.[191] Barth's reluctance may be a theological equivalent of Heisenberg's uncertainty principle which is generally accepted to define limits to achievable human knowledge in the physical world. Heisenberg, in 1927, showed that there are limits to what we can know in nature. There is a limit for example to how much can be known about atomic particles. The more exactly position is determined the less exactly momentum can be determined and vice versa. Barth was concerned that the more tightly Christians try to define how the Holy Spirit works the more likely they are to lose the life of the Holy Spirit's action. It may be that Barth's reluctance stands as a warning to further development of Pneumatology rather than a simple end to such exploration. Similarly Heisenberg's uncertainty principle did not spell the end for the development of particle physics.

Barth's Pneumatology does remain unfinished because ultimately Pneumatology cannot be completed in stand alone terms. Nevertheless, more needs to be said about the agency of the Holy Spirit while heeding Barth's warning not to lose the mystery. The proposed incarnational description of divine agency in humans extricates implications of perfect-being theology and the existence of the soul from the question of God's direct interaction in the world and thus from the dialogue between theology and science. While Barth's warning must be seriously noted, it is necessary to go beyond where he would permit himself to go in order to resolve these issues. The danger that Barth refers to in the defining of the action of the Holy Spirit is, however, no less a danger in every aspect of theology. In all areas of theology, the danger is that when its explanation is both comprehensive and clear, then human nature seeks to depend on the explanation rather than on the God who offers the life-giving relationship.

Barth, Incarnational Divine Agency and Resolving One Area of Tension

Whatever themes can be identified in Barth, and there are many, these ideas which affect his Pneumatology and this consequent discussion on the agency of the Holy Spirit in humanity are present in his theology from an early stage in his thought. However, what has been shown is of significance in that his pneumatological language changes little from his early lectures to his late *Evangelical Theology*. His use of the concept of *aufhebung*, the work of carefully and exhaustively re-examining logical assumptions in theology and secular philosophies, continues throughout his work and is one of the main reasons for the length of *Church Dogmatics*. This element

191. Ibid., 54–58.

of Barth's theological method is significant to the current discussion. It is precisely because the unexamined non-christological re-use of conclusions made by Augustine and early modern perfect-being theology and the two books metaphor has led to problems. Having examined Barth's recasting of Pneumatology as a result of his revision of assumptions leads to a set of conclusions about anthropology and divine interaction that are similar to those that have already been implicated in posing incarnational divine agency.

A good summary of Barth's view of the work of the Holy Spirit is that the Holy Spirit makes Christ known in

> His witness, in the mystery and miracle, the outpouring and receiving, of the gift of the Holy Spirit. He is the lighting of the light in virtue of which it is seen as light. He is the doctor *veritatis*. He is the finger of God which opens blind eyes and deaf ears for the truth, which quickens dead hearts by and for the truth, which causes the reason of man, so concerned about its limitations and so proud within those limitations, to receive the truth notwithstanding its limitations. He creates the Christian community, and in it the faith and love and hope of Christians, and in and with their faith and love and hope the knowledge of Jesus Christ as the One He is: the true Son of God who became and is also the true Son of man. He causes the apostles to know Him. He was the convincing power of their witness as it was heard and given again in the Church."[192]

The legacy of Darwin and Huxley's rejection of Paleyan perfect harmony and the metaphysical soul has developed into an apparently limited range of choices for the dialogue between theology and science. Apparently, either theology needs revision or God's perfection must be limited or the faith rejected. It is argued that this limited range of choices does not follow if divine agency can be expressed independently of perfect-being theology and Augustinian *ekstasis*. In early modernity these have been reappropriated in a manner that it was thought would guarantee that faith was based on rational evidence.[193] Darwin and many in his time held the assumption that the alternative to a Biblicism basing faith on rational evidence was to be found in experientially in the religious sentiments. Doubting the former and unsure of the later, Darwin and Huxley were led to express their inability to resolve their dichotomy as agnosticism.

---

192. CD IV/2:126.
193. Refer to section 2.5

Moving Away from Apparently Limited Choices

The relevant element in Barth's methodology pointing to the resolution of this underlying issue is to re-establish the beginning of theology with Christ, irrespective of where one personally begins or what one thinks one knows. In so doing additional choices can be offered to the apparently limited range of choices which arose between absolute certainty of Biblicist truth on one hand and experiential religion on the other. The understanding of divine agency in the world developed in early modernity from the understanding of divine agency in humans contained in the Augustinian description of inspiration in conjunction with late medieval perfect-being theology and application of the two books metaphor. What has been demonstrated at this point is that the presuppositions of divine agency in humans in Augustinian description of inspiration are neither adequate nor the sole way to adequately describe such agency. Divine agency in humans can be described without necessarily implying perfect divine action or an anachronistic metaphysical anatomy. Therefore, the precondition of the logic train highlighted earlier is not exclusive and the subsequent chain of logic is false. Furthermore, the other steps also do not follow. It can be concluded that: if God acts, God acts as God wills, and; if there are laws in nature that are consistently perfect this might but is not guaranteed to indicate something about God.

What has been argued is that resolving one particular unresolved underlying issue in the debate between theology and science can be achieved using the incarnational description of divine agency grounded in a pneumatological Christology of the incarnation. It has been argued that such grounding of divine agency in the nature of who God has revealed God's self to be avoids the insoluble problems that arise from starting with and trying to harmonise all of the perfections of God's freedom. The incarnation is in contrast the quintessential expression of the perfections of God's love.[194]

Recasting divine agency in humanity as constituted by the incarnation has not directly addressed the complex debates and issues related to perfect-being theology. It argues that the question of divine agency can be addressed and resolved separately. Also this discussion of divine agency has not addressed the two books metaphor which may well be exhausted. However, even when the notion of the two books has been abandoned that has not in itself changed the existence of the problem of divine agency in the world, as illustrated by Clayton. While both nature and Scripture may continue to be opened by correct interpretation and while both may indeed be inspired, there can be no assumption of similarity based in their mutual

---

194. Rom 5:8.

perfection. Further, if there is perfection of God's action it is not the fixed perfection that the natural theologians of the eighteenth century assumed it to be—perfect suitability ruling out the need to adapt.

There remains the possibility that divine agency in nature may be similar to that in humanity. Paul seems to think so in Romans 8 where the action of the Holy Spirit in the believer he describes as similar to that of creation in awaiting the revelation of the incarnate one. Nonetheless, the similitude is vaguely suggested and may be far from the close identity identified with the description of divine agency developed in early modernity.

Revising One Contributing Factor

One significant idea that the recasting of divine agency in humans directly affects is how the doctrine of inspiration should be reconsidered. Hence it is appropriate to reflect how elements of Augustine's description of inspiration are better addressed by the incarnational divine agency. Also the conversation with Barth's Pneumatology has suggested where the claims of what is a plausible and coherent proposal are stronger and broader than proposed. In addition, the proposal can be used to carefully extend Barth's Pneumatology just as Barth's Pneumatology suggests extension to the proposal.

1. The Augustinian *ekstasis* description depends on a radical dualism of soul or spirit from the physical, with the image of God reflected in an impassable element of the human spirit, through which the Holy Spirit must work. In contrast, the incarnational description does not presuppose that nature is dualistic, separated between the material and the spiritual. It rather merely assumes that the world is something in which God chooses to act. Barth's Pneumatology enjoins that there need be no radical dualism of soul and spirit and that God's action sustains human existence in Christ. This is a general theological principle not restricted to questions of Pneumatology or inspiration.

2. The Augustinian *ekstasis* description assumes a metaphysical anatomy that assigns functions of reason, judgement and direction to the soul which uses the senses and memory and directs the physical body. Barth's Pneumatology argues that the Holy Spirit works the work of Christ in any human actions and is not limited by any human state of consciousness or sanctity. The incarnational description of inspiration, further, does not assume metaphysics or a metaphysical anatomy and does not assume radical dualism between the human soul or spirit and the physical. Rather, the agency of the Holy Spirit in humans is

constituted by Christ's reception of the Holy Spirit in his enhypostatic humanity. The Holy Spirit is not limited to working through an impassable element of humans. Rather the Holy Spirit works on the whole person.

3. The Augustinian *ekstasis* description is built on a foundation which is a synthesis of Aristotelian philosophy, Neo-Platonism and classical medicine, not Scripture. Incarnational divine agency in humans does not depend on philosophical, scientific or medical ideas. Rather, it depends on the central mystery of the incarnation—how God can be human. Barth argues that in general, theology must remain independent of how the world's nature or purpose is understood. What the incarnational description does in one case, Barth claims, must be done generally. Barth, by considering the question of whether humans should be considered to be firstly individuals or community points out that theology must also ultimately be independent of anthropological assumptions drawn from the sciences. Humanity is redefined in Christ by his negation and reconstitution of its nature.

4. The Augustinian description assumes that the inspiring action of the Holy Spirit within the human person creates an *ekstasis* state similar to sleep, extreme fear, and death. *Ekstasis* is not necessary to the incarnational description of inspiration. Barth concludes that *ekstasis* cannot be automatic. The Holy Spirit does not work by moving the soul of the person to one side but rather by maintaining and enhancing the nature of the person as an individual or part of a community. The incarnational description points to the Holy Spirit's work in humans to reconstitute humanity to reflect and share in the enhypostatic humanity of Christ.

5. Augustine concludes that the more complete the *ekstasis* the more reliable the inspired action. The incarnational description of inspiration simply does not link possible *ekstasis* to reliability. For Barth there cannot be differences in degree of inspiration measured by the results of the action. Barth specifically argues that humans have no guarantee of the accuracy or perfection of inspired action; rather they must depend on God.

6. The Augustinian *ekstasis* description of inspiration assumes that the most complete state of *ekstasis* is totally reliable and that all Scripture exemplifies this degree of *ekstasis*. Barth argues that the only guarantor of revelation or even of Scripture is God. With incarnational divine agency, inspiration in humans depends on the Holy Spirit's action but

remains fully human. While not denying the possibility of perfected human action as a result, the reality or even the effectiveness of such inspiration can occur in spite of any human limitations including error.

7. The Augustinian *ekstasis* description of inspiration concludes that Scripture must be reliable and must have been written in a special complete state of *ekstasis*. The incarnational description treats the writing of Scripture as being in the same class as other activity conducted under the inspiration of the Holy Spirit. Barth concludes that Scripture reliably achieves what God wants it to achieve, God alone giving it finally authority. While this does not rule out the infallibility of God's action in inspiration, such an expression of perfection will not meet any definition humans independently define. Scriptural inspiration does not follow as a special case. For Barth, infallibility becomes a non-issue.

The incarnational description of inspiration cautiously adds detail to the description of how the Holy Spirit works particularly in relation to humans in the world. It goes beyond where Barth stops but does not try to resolve the central mystery of the Christian faith—how God became human and can be both divine and human. Incarnational divine agency in humans as applied to inspiration is distinguished by the following four features:

1. It anchors God's action in the central mystery of the Christian faith independent of any cosmology.
2. It anchors inspiration by describing that the agency of the Holy Spirit in humans derives from Christ's reception of the Holy Spirit into his humanity and is applied by Christ's action.
3. It describes inspiration independently of anthropology, neurochemistry and psychology. This leaves theology free to engage developments in these fields with or without metaphysics as seems appropriate.
4. Scriptural inspiration becomes one instance of the Holy Spirit's inspiring work rather than a special case. It may be considered canon, thus becoming the measure for assessing other instances of the Holy Spirit's inspiring work, similar to ways in which Scripture acts as the measure for Christian belief, ethics and worship.

It is important to note Barth's self-limitation in relation to this discussion. He did not and would not have supported the detail of incarnational divine agency in humans of inspiration. However, by dissolving and recasting the apparent dichotomy between perfection and experience incarnational divine agency does offer a way forward for the debate between theology and science. In demonstrating that incarnational divine agency in humans is a plausible and coherent alternative to the Augustinian description, a significant stumbling block to the dialogue between theology and science is removed.

To this point the recasting of divine agency has been in relation to divine agency in humans. As observed in the third chapter, Newton historically drew a parallel between divine agency in humans and divine agency in the putative soul of the world. However, no such parallel is logically essential. Nevertheless, if such a parallel were to be drawn afresh, rather than postulating that the natural world would reflect God's perfect-being, it might be suggested that the natural order expresses the nature and purposes of Christ expressed in his incarnation: that God is active in the world; that Christ created and sustains and renews the world in himself, and; that this is an expression of God's love whether or not it is recognized as such. Such christological understanding of creation has an expected echo in Scripture. "In him was life, and that life was the light of all people . . . He was in the world, and though the world was made through him, the world did not recognize him."[195] Such a parallel could point to the cosmic extent of redemption in Christ.

To sound a note of caution, it is argued that Barth's concerns regarding the relationship of theology to understandings of the natural world should be taken seriously. While such a recast understanding of divine agency in the world might suggest new ways to understand the world, it must also stand apart with the freedom to reject whatever system of thought or philosophy is constructed on it. Such a parallel might be argued to guarantee that creation has a purpose but not guarantee that the purpose be reflected in nature or indeed that such purposes be intelligible. It would not offer a theological guarantee for the universal consistency of the laws of nature.

At this point, the case has been made for the consideration of the incarnational description of divine agency as it applies to the personal nature of divine contact with humans. In freeing consideration of divine agency within humans from necessarily presupposing divine perfect action in the world and the notion of a metaphysical soul, it could also be argued that these presuppositions need not apply to the broader question of divine agency in the world or creation in general.

---

195. John 1:4, 10 (TNIV).

CHAPTER 6

Dialogue with One Obstacle Removed

IN THE LIGHT OF the detailed discussion of this book, how might we speculate that divine agency in the world might be described? One possible theological approach may be to postulate that such general divine agency may parallel the incarnational description. Could it be that by *enhypostatically* preserving the full humanity of Christ which has been argued becomes constitutive for the Holy Spirit's interaction with humanity, also extends to this same preservation in including the matter constituting Christ's resurrection body? This might imply that the *enhypostatic* way in which God interacts with matter and the laws governing those atoms and their interaction may be constitutive for divine creation and interaction in the world. Such a Trinitarian and christological understanding would not be inconsistent with Christian notions of the creation. That is, "all things came into being through him."[1] Further the cosmological scope of redemption referred to by Paul in, the liberation of creation from decay is somehow "in the same way as the Spirit helps us in our weaknesses."[2] The reversal of the decay inherent in the universe in the person of Christ may also offer hope for the ultimate fate of the universe other than the rather pessimistic heat death scenario offered by cosmology in response to the burden of the laws of thermodynamics. This however can only remain

1. John 1:3
2. Rom 8:21–22 and 26.

speculation, unless perhaps, it became possible to demonstrate changes to the laws of physics or particular fundamental physical constants.

The incarnational description of divine agency is coherent and plausible warranting serious consideration. The question remains what implications this may have: for the debate between theology and science; for each discipline; for doctrine, and the shape of academic debate.

Discussion of God's immediate action in the world need not be predicated by such notions as: all things having to have a created purpose; God always having a motive for acting; God always acting for the best; the world being the best of all possible worlds, and; the existence of the soul. The incarnational description of divine agency removes the insistence that such agency necessarily reflects the attributes of perfect-being. This removes Darwin's major issue with Paley's proof of faith based on perfectly created divine harmony in the world. By not assuming any form of metaphysics, incarnational inspiration also frees theological discussion from anthropology. Huxley's question about the location of the soul, if it exists, becomes a non-issue. The existence or non-existence of the soul is of no consequence to this description. God interacts with whatever it is that humans actually are.

Theological questions about providence and order in nature remain, but these discussions are no longer hamstrung by presuppositions about what God must be and what God must do. Why the world follows laws, is an issue science cannot answer except to say that is how it is. Why? is a legitimate question for theology to pursue. Freed of necessarily perfect divine action, theologians may find some unexpected answers. Providence has the opportunity of becoming a richer and more complex issue. It may well be asked how it is that God uses what seems to be chance, randomness and directionless chaos to bring about order and beauty while directing the world toward God's chosen ends.

The proposed incarnational description removes from the interaction between traditional Christian theology and the sciences, assumptions which have been problematic in the debate in the last century and a half. This is not by way of theology's accommodation with science but rather as the result of a careful re-examination of that which in theology were undistinguished flaws in the traditional explanation of the truth about how God interacts with the world and with humans. This re-examination of assumptions is one tool which can enable a corrected or revised traditional theology to interact with another academic field on its own terms.

Revision of divine agency as proposed also has implications for theology as well as science. While assumptions and descriptions may shape theological debate, theology is ultimately not the sum of its descriptions.

The God to whom Christian theology seeks to bear witness and through faith to understand is not bound to these descriptions. Good theology describes God well, whereas bad theology obscures or creates obstacles to this knowledge. Incarnational divine agency might be described in a way that sits within Lindbeck's proposal for developing rule theory in contrast to propositional statements or Experientialism/Expressivism. It may be a way to express an unconditional and permanently essential doctrine in Lindbeck's taxonomy. That is, the agency by which Holy Spirit acts in the world and through humans is constituted in the hypostatic union of the person of Christ. The anthropological, anatomical and metaphysical details become conditional applications of the essential doctrine, interacting with how humans are best understood.

The non-christological generic understanding of divine agency in the world has doubtless helped to shape both post-Reformation theology and the development of science. Nonetheless, it also underlies some of the most persistent stumbling blocks to dialogue between the disciplines. It should not be surprising to find that there have been flow-on influences in the disciplines of academic thought which have been shaped by theology and science.

Revised Divine Agency and the Dialogue with Science

That the incarnational description of divine agency in humans warrants serious consideration is a necessary step in reframing the dialogue between theology and science. It is necessary but in itself insufficient to reorder the whole dialogue. This removes only one of the unresolved underlying issues over which the dialogue has repeatedly stumbled. By revision of one important theological idea, hope is offered in answer to Bowler's pessimism as to whether resolutions can be found to renew the whole dialogue. The appropriate shape of such a renewed dialogue is an open question, whether the dialogue between theology and science should be one:

1. that is a partnership of interested travellers;
2. where one discipline defers to the other;
3. where each provides a source of mutual revision to the other;
4. in which they continue to humbly listen to each other;
5. in which they mutually support each other.

Whatever the appropriate shape of the dialogue in the future, it cannot be a full dialogue unless both disciplines can engage with each other while maintaining their integrity as disciplines. Theology depending on the non-christological generic description of divine agency in the world has not achieved this. This has led to the apparent impasse between Biblicism and Experientialism. The incarnational revision to divine agency does not lead to the same impasse. Theology revised to depend on an incarnational understanding of divine agency may offer a way for theology to maintain its integrity freed from presuppositions that are grounded in another field of study.

Continuing interaction with science may and in all likelihood will continue to challenge theology. It is to be hoped that challenges will result in theology amending itself with revisions developed from the grounds of its own methodologies. That these revisions will help it to more closely describe truth in relation to God. This is in contrast to radical revision based in ideas native to science. In the past, such radical revision has resulted in forms of theology that many Christians cannot own as describing their faith.

Also revising divine agency christologically may have implications for revising theology. Not only have the non-christological and generic assumptions related to perfect-being theology and Augustinian metaphysics been buried deep in the foundations of the development of the dialogue between theology and science, they have remained influential in the development of theology. By highlighting the problems with these assumptions, it is to be hoped that incarnational divine agency might contribute to theology's revision of itself as a discipline.

What is to be hoped is that application of this important revision will contribute to theology being able to provide to itself a satisfactory answer to Lash's observation that the serious engagement with the dialogue between theology and science is a matter of truthfulness "integral not only to morality but to sanity"[3] and that such revision of theology will provide a Christian seriously engaging with the claims of science confidence about "likely or appropriate forms of survival (if any) of religious belief and practice."[4] The existence of a coherent plausible description of divine agency means that the eighteenth-century understanding of divine agency in world is neither all that can be said about the subject nor even an adequate summary. God acts as God wishes. Christian theology can continue in dialogue noting that the predicates of divine agency in the non-christological and generic assumptions related to perfect-being theology and Augustinian metaphysics

---

3. Lash, *Believing Three Ways in One God*, 10.
4. Ibid.

are significant, albeit, embarrassing facts of theological history. However, this is not simply an embarrassment to theology, science too is included. Science would not have grown to the extent it has without having being built on these same mistaken assumptions.

As science has evolved from its roots in seventeenth-century natural philosophy, it has progressed by leaving alone "hypothetical" questions and concentrating on the "how" of the workings of nature. Newbigin, in citing Bacon's and later Newton's dictum "I do not hypothesise" in relation to the development of science, makes the useful observation that there has been a largely unrealized sacrifice as a result. His persuasive observation is that modern science has no inherent method for dealing with purpose, God and ethical questions. Purpose and divine influence address answers to the "why questions." This, however, has not stopped science being used to address such issues, with ambiguous results.[5] The persistence of unresolved underlying issues between theology and science has eroded confidence from a scientific perspective that theology has authority to make truth claims regarding ethics or purpose. In dealing with a significant underlying issue, incarnational divine agency may help to stem the erosion.

Science's inability to deal with "why?" questions has also eroded confidence that science alone can deal with moral dilemmas arising from "scientific advances" in medicine, society, warfare and the environment. When science is forced to venture into these "why question" areas, there are unexamined presuppositions needing careful examination. There is a need, as Watts has indicated, to examine carefully the presuppositions of both the scientist and the theologian.[6] Jaki sees as absolutely essential the need to critically re-evaluate philosophical presuppositions for discussing truth and ethics and has no doubt that this examination is vital in ongoing academic debate. He is, however, unenthusiastic regarding leading scientists' demonstrated ability to do this quoting Einstein's self-depreciating dictum, "the man of science is a poor philosopher."[7] If the interaction between theology and science can avoid antagonism or agnosticism arising from unrealized untenable assumptions, then the possibility of a broader mutually beneficial or even synergistic interaction may be possible. The proposed revision to divine agency may open a door such re-evaluation that is not based in obsolete or contested assumptions. This revision was developed from a careful re-examination of firmly and long held theological assumptions. I argue

---

5. Newbigin, *The Gospel in a Pluralist Society*, 8.
6. Watts, "Are Science and Religion in Conflict?," 136–37.
7. Jaki, *Means to Message: A Treatise on Truth*, 43–61.

that science, too, can benefit from similar careful re-examination of its own firmly and long held assumptions.

While, the incarnational divine agency shows it is possible to break free on a key unresolved issue, this does not automatically ensure that a broad consensus and mutual constructive interaction will occur in all areas. A common failing of works giving a broad overview of the theology and science dialogue is that they assume that an over simplified answer to the relationship is possible, be it harmony, segregation or conflict. This fails to give adequate recognition to the complexities and fluidity of theories within specific parts of theology, science and philosophy. There must remain the possibility of irreconcilable ideas and theories. For example (even if extremes like creation science, on one side and the rhetoric of the militant atheists on the other are excluded) in relation to evolutionary theory, it seems likely that there will be continuing disagreement between aspects of conventional Christian theology and scientific theory. There may well continue to be internal inconsistency in science or theology. For example, Hubble's first estimate of the ages of the universe according to his Big Bang theory was 1.5 billion years compared to the generally accepted geological age of the Earth at 4.5 billion years. This example though, is still an area of ongoing debate. Hubble's estimate has been revised upward to be in excess of 13.5 billion years using alternate dating methods.[8] It may be possible and is indeed likely that even if mutually beneficial dialogue is possible for key issues, there will be degrees of continued antagonism and agnosticism between aspects of both fields. Nonetheless, as Watts states, "Perhaps the key requirement, if there is to be fruitful dialogue between theology and science, is mutual respect between the methods and epistemologies of theology and science. That is not something that currently obtains. On the contrary there is considerable methodological suspicion."[9]

In the present dialogue, mutual quotations demonstrating incisive understanding of the other discipline have been rare, unlike debate in the seventeenth and eighteenth centuries. There are exceptions: for example the respective books of Torrance and Davies.[10] Together these two books constitute less than a full debate. Largely, academic theology and science are alien to each. This, however, is through ignorance rather than of necessity. As Barth claimed, however, the alienness of the other discipline can be recognized without preventing the two disciplines from working together. While

---

8. Davies, "Physics and the Mind of God: The Templeton Prize Address," 132–34.

9. Watts, "Are Science and Religion in Conflict?," 136.

10. Davies, *Runaway Universe*, cited in Torrance, *Divine and Contingent Order*, 56n; Torrance, *Divine and Contingent Order*, cited in Davies, *The Mind of God*, 167–72.

mutual awareness improves this dialogue, mutual benefit does not require that theology and science agree. Mutual challenge and cross-fertilization of ideas require freedom to be different.

The proposed revision to divine agency is one essential step in making renewed dialogue possible without compromising either science or Christian theology. In the past dialogue between theology and science is not well served by revision of Christian theology in the light of "scientific advance." As Smart stated, "in the Middle Ages theology was the Queen of the Sciences, while in modern times religious studies has become the Knave of Arts."[11] Such compromising revision of theology makes the initial assumption that theology is the knave to the regal status of Science. The knave will inevitably, he argues, be enslaved to a mistress who has little or no concern for the truth theology confesses. It is to be hoped that a different and more notably christological theology of divine agency might contribute to both the foundations of contemporary debate and to establishing a firmer place for theology as an academic discipline in the secular academy.

The application of an incarnational understanding of divine agency would be one necessary step in freeing Christian theology from presuppositions that are grounded in another field of study. This would enable theological truth to stand on its own, grounded in Jesus Christ, enabling it to interact as a fully-fledged discipline in its own right with related fields of academic endeavor rather than merely the barely tolerated elderly spinster aunt of contemporary academic and moral debate. If not the Queen of sciences, theology may again become part of the academic royal court rather than being merely Smart's "Knave of Arts."

Revised Divine Agency and Doctrine's Function

The question may also be asked about how the incarnational revision divine agency may influence theology. The place and function of doctrine within Christian theology continues to attract significant debate. The present discussion on inspiration cannot be divorced from considerations of this debate. Lindbeck has proposed a rule-based cultural-linguistic model for describing the functioning of doctrine as an answer to the discussion of doctrine becoming polarized between propositional statements and descriptions of Experientialism/Expressivism. This polarisation has led, Lindbeck asserts, to despair about the place of doctrine in theology. The polarisation parallels the apparent dichotomous choice, which developed in the nineteenth century, between basing theology in Biblicism or the

11. Smart, *Secular Education and the Logic of Religion*, 4.

religious affections. That dichotomy has been shown to have grown out of the legacy of the assumptions underlying divine agency, perfect-being theology and the two books metaphor. In arguing for the serious consideration of the incarnational divine agency this dichotomy has been shown to be false. There are, however, significant points of contact between the false dichotomy of choice between Biblicism and the religious affections on the one hand and the polarisation between propositionalism and Experientialism/Expressivism related to doctrine on the other. Biblicism, for example, has been commonly propositional in its descriptions of doctrine.

It might be possible to deal with divine agency and inspiration within the Lindbeck taxonomy of doctrines. "Now if doctrines that propose beliefs are treated as rules . . . [t]hey also can be viewed as unconditionally or conditionally necessary, as permanent or temporary, as reversible or irreversible."[12] Incarnational divine agency may be placed as expressing an unconditional and permanently essential doctrine in Lindbeck's terms. That is, the Holy Spirit acts in the world and through humans deriving from the person of Christ. The anthropological, anatomical and metaphysical details become conditional. Their description interacts with how humans are best understood at particular places and times. In that respect inspiration based on such agency also becomes a conditional doctrine more concerned with theological anthropology and Pneumatology than part of the doctrine of Scripture. Francis Watson's challenge[13] as to whether inspiration as a doctrine has a place can be answered by affirming that it does as part of general Pneumatology and theological anthropology, though not of necessity as unconditional and permanently essential aspect of the doctrine of Scripture.

Nonetheless, Lindbeck's taxonomy of doctrines suffers from the assumption of modernity's definitions. The way Lindbeck frames his cultural-linguistic proposal depends in part on issues intertwined with those of perfect-being theology and Augustinian metaphysics. The notions of doctrine, being considered as proposition vs. the expression of experience makes similar kinds of distinctions in the authority of doctrine as identified as developed with the false dichotomy the non-christological generic understanding of divine agency developed in making choice for the authority of faith between scientific fact and religious feeling. However useful Lindbeck's taxonomy may be as a tool for describing the use of doctrine within the Christian community, it suffers from presuming the very matters

---

12. Lindbeck, *The Nature of Doctrine: Religion and Theology in a Postliberal Age*, 8.

13. Watson, "Hermeneutics and the Doctrine of Scripture: Why They Need Each Other," 9n20.

entwined with assumptions that have proven problematic with consideration of divine agency.

If some of the arguments from Barth's theology which have helped to support the coherence and plausibility of incarnation are to be taken seriously then aspects of Lindbeck's theory of doctrines must be questioned. In particular, Lindbeck's assertion of the primacy of a religion as "a kind of cultural and/or linguistic framework or medium that shapes the entirety of life and thought"[14] as a precondition to understanding theological/doctrinal truth claims must be challenged. Hunsinger has summarized Lindbeck's proposal thus, "[t]he coherence of linguistic usage with behavior in accord with communal norms (rightness) is a condition for the possibility of using a sentence to refer accurately to the external domain of ultimate reality (truth)."[15] Not only would Barth decry that the alienness of such a presumed philosophy being a precondition for theology, he would also deny the appropriateness of presuming a theory of culture and language as necessary to enable God to be understood. Hunsinger, in summarising Barth, has also stated,

> [t]heology ought never to be pursued or presented by the Christian community, he believed, as though the reality of the living God could be the object of neutral or detached consideration. . . . The truth of the gospel, as Barth understood it, was not only entirely self-involving from the human side, but also and primarily from the divine side.[16]

Treating doctrine as culturally or linguistically conditioned, while avoiding propositionalism or experientialism still fails, because it treats doctrine as a concept detached from God. "The cognitive truth of a theological assertion does not finally depend on the rightness of the community's (or the individual's) performance in a correlative form of life. Rightness and truthfulness are by no means irrelevant to the valid assertion of the truth, but neither are they the final and overriding conditions for its possibility."[17] The truth of a doctrinal statement depends firstly on God's self-involved commitment to its expression. The expression of doctrine is itself an outworking of the divine agency in the human community of the church. It is, as Barth suggested, the ongoing in-breathing work of the Holy Spirit and so is a continuing reflection of the Holy Spirit's agency. By applying the christo-

---

14. Lindbeck, *The Nature of Doctrine: Religion and Theology in a Postliberal Age*, 33.

15. Hunsinger, "Truth as Self-Involving: Barth and Lindbeck on the Cognitive and Performative Aspects of Truth in Theological Discourse," 44.

16. Ibid., 41.

17. Ibid., 49

logical notion of divine agency proposed, inspiration should be considered broadly in Pneumatology and not as part or pillar of the doctrine of Scripture. Inspiration as a doctrine acts as a signpost to relationship to help a person from where they are now to a closer relationship with God. Better understanding of the nature of God and of that relationship can be part of that drawing closer but is not a precondition of the relationship.

Revision of the doctrine of inspiration as proposed has implications for theology, *per se*, as well as for the dialogue between theology and science. As a result of the discussion in this book Lindbeck can be seen as not completely avoiding some of the consequences that have developed as a result of the assumptions of Augustinian *ekstasis* inspiration. This may in turn be illustrative of the difficulties of untangling theology from these long assumed but erroneous assumptions. The renewal of dialogue between theology and science that may result from the resolution of these underlying issues is, in turn, not unrelated to the renewal of the discussion of doctrine within theology. Because these assumptions have lain unexamined for so long there are likely to be many flow on affects in academic disciplines. In relation to the theology and science dialogue this revision of the expression of the doctrines of divine agency and inspiration may help to sway academics to engage in a deeper interaction than previously possible. In particular it may allow for a new level of interaction between a more traditional Christian theology and the sciences regarding public truth claims. Many such interactions have been characterized by a radical revision of Christian traditional teaching or have reached the impasse Peter Bowler has identified as having occurred a number of times since the end of the nineteenth century.

Implications for the Current Debate: Foundations of Shifting Sand: Which Assumptions? Whose Analysis?

The non-christological generic understanding of divine agency in the world and *ekstasis* description for inspiration have played an important role in the development of the conventions of thought that govern formal academic debate, particularly as they relate to making broadly defensible public truth claims. Weaknesses in these descriptions have led some to be disillusioned in relation to finding truth. Whether public truth claims can be persuasive relates to the questions: "How are logic and rhetoric balanced in debate?" and "How should logic and rhetoric be managed?" These are two separate and important questions that are not unrelated to the theme of this book.

This study has implication for the nature of the conduct of debate in academic argument. In the period between the early 1700's and the 1800's the very methods and reasons for investigation, rationalising and explaining the world changed. Harrison has argued, and my discussion agrees, that the revival of Augustinian thought has been an important influence on the development of modern science as an academic discipline. While the changes in attitude to the doctrine of divine inspiration (in its Augustinian form) are often mentioned and even described, there has been no serious study of the implications and influences of its particular presuppositions upon the development of modern thought and its contribution to the understanding of divine agency in the world. While the influence of perfect-being theology and the notion of the two books of God's revelation have been discussed at length, such discussions have not adequately identified the shape of divine agency widely understood throughout the eighteenth and early nineteenth centuries. As shown, there are inherent theological weaknesses in those presuppositions and the understanding of divine in the world which developed. One might wonder what difference would changing the presuppositions have made to the development of modern thought and the marginalisation of theology within mainstream academic debate? If such a misconception of divine agency has been one of the important factors in determining the rise of modern academic debate, then the methods that would have been used to develop this type of discussion may need to be revised. This might lead to despair that an adequate starting point could be found to begin to address a self-referential topic in such a way that the methodology, rationality and even the reason for selecting the topic may on later consideration become irrelevant. As with all learning, however, there are no guarantees. The process is begun in hope and humility, with a faith that better understanding is possible.

The development of modern thought has led to a complex three way interaction between philosophy, theology and science. Most current debate assumes a strict implied precedence of philosophy over science and science over theology. Just as most scientists have historical amnesia about the theological roots and philosophical implications of their discipline, the same could be said for both philosophy and theology.

A problem with contemporary dialogue between theology and science has been its historical illiteracy. In addition, there has been simultaneous assumption of the prejudices, ideologies and narratives that have shaped the development of these fields of study in their forms in the modern period. History offers a different perspective to assess both the weaknesses and strengths of these other disciplines. There is an old maxim: "Those who cannot remember the past are condemned to repeat it."[18] As Bowler convincingly shows in his study of the interaction between theology and

18. Santayana, "The Concise Columbia Dictionary of Quotations."

evolutionary theory during the twentieth century, there is ample evidence of the truth of this maxim when similar debates on similar issues with similar solutions and impasses have arisen almost once a generation since the turn of the twentieth century.[19]

Similarly, awareness of the limitations of historiography will help to avoid some of the glaring errors of presentism, Whiggism and heroism that have been often identified in histories of "Science and Religion." Not uncommonly theologians, scientists and philosophers make assumptions which come unexamined from postulates that have their grounding in schools of one of the other disciplines. The danger is that if some philosophical corollary, scientific theory or doctrinal statement is first assumed without critical evaluation, then all the conclusions built upon that assumption may well be wrong.

Specifically, the concern here has been that this is exactly what has happened with divine agency described in terms of the generic impersonalized assumptions that developed out of the conjunction of perfect-being theology, the two books and *ekstasis* inspiration. When these assumptions are recast, then much else needs to be and can at last be reworked. As much of the interaction between theology, science and philosophy has taken old assumptions for granted and thus been naïve, it should be no surprise to find further inconsistencies and disagreements having more to do with a lack of understanding of sources and the pedigrees of some cherished or "obviously true" assumptions in each field. Only by diligently having the courage to work things out in the detail of each discipline's assumptions can the dialogue between theology and science surmount what Lash calls the enlightenment legacy of the "crisis of docility."[20] The unattractive alternative is remaining inhibited and enslaved by "meanings and values, descriptions and instructions, imposed by other people, feeding other people's power."[21] To be so inhibited and enslaved is to be held by the demands of often incompatible and contradictory meanings and values, descriptions and instructions.

There is a need too for awareness about the limitations of theories and assumptions within each discipline. The need for better understanding of intra and inter disciplinary limits is evident. The hope is that fruitful debate will not require the impossible goal that the debaters have expertise in every discipline.

19. Bowler, *Reconciling Science and Religion*.
20. Lash, *Believing Three Ways in One God*, 10.
21. Ibid.

What is proposed is not a simple solution to a complex problem, but rather a way to begin to readdress issues. With a new starting point in the incarnational description of divine agency, and using its implications as part of a different method, it may be possible to redevelop a range of contemporary issues in a way that does not lead to polarisation, as has been the case.

There are conventions to academic study which have their roots in these very same sorts of assumptions. If starting points together with methods and rationales find themselves moving on a shifting foundation, like sand, then where can there be hope that better understanding can be achieved? Which assumptions and whose analysis should be used? Where is hope that interdisciplinary dialogue will be possible let alone fruitful? There is no assurance and can be no guarantee that any starting point will be either optimum or "correct."

The utility of a starting point lies in whether it enables an answer to be found. A starting point is not, it is hoped, where the investigator will stand at the end. This flies in the face of the too common theological practice of arguing for the appropriateness of a chosen starting point. Lash's comment is both provocative and appropriate. The justification of a starting point is,

> a largely futile exercise because, if one thing is certain in this life, it is that none of us begins at the beginning. We find ourselves somewhere, discover something of what went before, of how things went in order to bring about the way they are. Growing up is largely a matter of learning to take bearings. A more fruitful question than "where should we begin?" would almost always be "Where, then, do we stand?"[22]

The hope is that the reformulation of divine agency will assist the theology and science dialogue answer where to stand by redefining the place of theology within academic debate as well as remove problems associated with the nature of seeking truth. That there are answers to the questions posed is a faith statement, a statement of hope. While there may be no guarantees of a result, there is a theological assumption that gives hope. The hope of the gospel is that no starting point can be too far away. For God does choose to be known, then God's action with humanity is in some way intelligible.

---

22. Ibid., 2.

Bibliography

Agassiz, Louis. "Upon Glaciers, Moraines, and Erratic Blocks." *Edinburgh New Philosophy Journal* (1838).
Aherne, C. "Commentaries on the Bible." In *Catholic Encyclopaedia*, edited by K. Knight, 4. New York: Appleton, 1908. http://www.newadvent.org/cathen/04157a.htm (accessed 14 March 2015).
Aland, Kurt, et al., eds. *The Greek New Testament*. Stuttgart: United Bible societies, 1983.
Alexander, H. G. *The Leibniz-Clarke Correspondence*. Manchester: Manchester University Press, 1956.
Alonso-Schökel, L. *The Inspired Word*. London: Burns & Oates, 1967.
Anderson, R. S. "Barth and a New Direction for Natural Theology." In *Theology Beyond Christendom*, edited by J. Thompson, 241–66. Allison Park, PA: Pickwick, 1986.
Appleby, J., L. Hunt, and M. Jacob. *Telling the Truth about History*. London: Norton, 1994.
Aquinas, T. *Summa Theologiae*. Translated by the Fathers of the English Dominican Province. Benziger Bros. ed. 1947. Reprint, New York: Christian Classics, 1981.
Arbib, M. "Toward a Neuroscience of the Person." In *Neuroscience and the Person: Scientific Perspectives on Divine Action*, edited by R. J. Russell et al., 77–100. Vatican: Vatican Observatory and the Center for Theology and the Natural Sciences, 1999.
Ashton, J. F., ed. *In Six Days*. Sydney: New Holland, 1999.
Athenagoras. *A Plea for the Christians*. In *Ante-Nicene Fathers*, edited by Roberts, A. and Donaldson, J. 2:256–99. Edinburgh: T. & T. Clark, 1994.
Augustine. *Contra Epistolam Manicaei Quam Vocant Fundamentum*. In *Post Nicene Fathers Series 1*, edited by P. Schaff, 4:220–72. Edinburgh: T. & T. Clark, 1996.
———. *De Anima et Eius Origine*. In *Post Nicene Fathers Series 1*, edited by P. Schaff, 5:792–938. Edinburgh: T. & T. Clark, 1996.
———. *De Bono Viduitatis*. In *Post Nicene Fathers Series 1*, edited by P. Schaff, 3:742–73. Edinburgh: T. & T. Clark, 1996.
———. *De Civatae Dei*. In *Post Nicene Fathers Series 1*, edited by P. Schaff, 2:8–1086. Edinburgh: T. & T. Clark, 1996.
———. *De Doctrina Christiana*. In *Post Nicene Fathers Series 1*, edited by P. Schaff, 2:1087–1262. Edinburgh: T. & T. Clark, 1996.

———. *De Genesi ad Litteram*. Ancient Christian Writers 41–42. New York: Paulist, 1982.
———. *Expositions of the Psalms*. In *Post Nicene Fathers Series 1*, edited by P. Schaff, 8:2–1485. Edinburgh: T. & T. Clark, 1996.
———. *First Epistle of John*. In *Post Nicene Fathers Series 1*, edited by P. Schaff, 7:917–1052. Edinburgh: T. & T. Clark, 1996.
———. *The Gospel of John*. In *Post Nicene Fathers Series 1*, edited by P. Schaff, 7:6–916. Edinburgh: T. & T. Clark, 1996.
———. *Letters*. In *Post Nicene Fathers Series 1*, edited by P. Schaff, 1:391–1207. Edinburgh: T. & T. Clark, 1996.
Aune, D. E. *Prophecy in Early Christianity and the Ancient Mediterranean World*. Grand Rapids: Eerdmans, 1983.
Bacon, F. *The Great Instauration*. Translated by J. Spedding, R. L Ellis., and D. Heath. Vol. 8, *The Works of Francis Bacon*. Boston: Taggard and Thompson, 1863.
———. *Novum Organum*. In *The Works of Francis Bacon*, edited by B. Montagu, 14:1–213. London: Pickering, 1620.
Baker, A. D. "Theology and the Crisis in Darwinism." *Modern Theology* 18, no. 2 (2002) 183–215
Baille, J. *The Idea of Revelation in Recent Thought*. New York: Columbia University Press, 1961.
Bainton, R. *Here I Stand*. London: Lion, 1978.
Barbour, I. G. *When Science Meets Religion*. San Francisco: HarperCollins, 2000.
———. "Evolution and Process Thought." *Theology and Science* 3, no. 2 (2005) 161–78.
Barker, E. "Science as Theology—the Theological Functioning of Western Science." In *The Sciences and Theology in the Twentieth-century*, edited by A. R. Peacocke, 262–80. Notre Dame, Indiana: University of Notre Dame Press, 1981.
Barnes, T. D. *Tertullian: A Historical and Literary Study*. Oxford: Clarendon, 1985.
Barr, J. *Escaping from Fundamentalism*. London: SCM, 1984.
———. *Fundamentalism*. London: SCM, 1961.
Barrett, Justin L. "Is the Spell Really Broken? Bio-Psychological Explanations of Religion and Theistic Belief." *Theology and Science* 5, no. 1 (2007) 72.
Barrett, P. H. "Darwin's Early and Unpublished Notebooks." In *Darwin on Man*, edited by H. E. Gruber, 259–480. London: Wildwood House, 1974.
Barth, K. *Anselm: Fides Quaerens Intellectum*. London: SCM, 1960.
———. *Church Dogmatics*. 4 vols. Edinburgh: T. & T. Clark, 1936–1969.
———. *Die Römerbrief*. 3rd ed. Zürich: TVZ, 1924.
———. *Dogmatics in Outline*. London: SCM, 1949.
———. *The Epistle to the Romans*. Translated by E. C. Hoskyns. Oxford: Oxford University Press, 1968.
———. *Evangelical Theology*: Weidenfeld and Nicolson, 1963.
———. *The Göttingen Dogmatics*. Vol. 1. Grand Rapids: Eerdmans, 1990.
———. *The Holy Spirit and the Christian Life*. Louisville: Westminster John Knox, 1993.
———. *How I Changed My Mind*. Edinburgh: Saint Andrew, 1966.
———. "Nein!" In *Natural Theology: Nature and Grace and No!*, edited by J. Ballie, 65–128. London: Centenary, 1946.
Barton, R. "An Influential Set of Chaps: The X-Club and Royal Society Politics 1864–85." *British Journal for the History of Science* 23 (1990) 53–81.

Bayly, B. *An Essay on Inspiration*. London: Wyat, 1708.
Benoit, P. *Revelation and Inspiration*. Chicago: Priory, 1960.
Bowler, P. J. *Charles Darwin: The Man and His Influence*. London: Blackwell, 1990.
———. "Development and Adaptation: Evolutionary Concepts in British Morphology." *British Journal for the History of Science* 12 (1987) 283–97.
———. *The Eclipse of Darwinism*. Baltimore: Johns Hopkins University Press, 1983.
———. "Evolution and the Eucharist: Bishop E. W. Barnes on Science and Religion in the 1920's and 1930's." *British Journal for the History of Science* 21 (1998) 453–67
———. *The Non-Darwinian Revolution: Reinterpreting a Historical Myth*. Baltimore: Johns Hopkins University Press, 1988.
———. *Reconciling Science and Religion: The Debate in Early Twentieth-century Britain*. Chicago: University of Chicago Press, 2001.
Brennan, R. "Has the Frog Human a Soul?" *Scottish Journal of Theology* 66, no. 4 (2014) 400–413.
———. "On Why We Should Agree with Contemporary Atheists—or Why A Generic God Does Not Exist." *Christian Perspectives on Science and Technology* 9 (2013) 1–10.
———. "Reason or Religious Affections—a False Dichotomy in Divine Revelation as a Legacy of the Early Modern Construction of Divine Agency in the World." *Christian Perspectives on Science and Technology* 9 (2013) 1–13.
Brewster, D. *Life of Sir Isaac Newton*. New York: J. & J. Harper, 1833.
———. *Memoirs of the Life, Writings, and Discoveries of Sir Isaac Newton*. 2 vols. Edinburgh: Constable, 1855.
The British Palladium. London: Stationer, 1758.
Brooke, J. "The Changing Relations between Science and Religion." In *Interdisciplinary Perspectives on Cosmology and Biological Evolution*, edited by L. A Trost, 3–18, Adelaide: Centre for Theology and the Natural Sciences, Berkley, 2001.
———. "Reading the Book of Nature." *MetaScience* 8, no. 3 (1999) 444–48.
———. *Science and Religion: Some Historical Perspectives*. Cambridge: Cambridge University Press, 1991.
Brooke, J., and Cantor, G. *Reconstructing Nature: The Engagement of Science and Religion*. Edinburgh: T. & T. Clark, 1998.
Brown, C. Mackenzie. "The Conflict between Religion and Science in Light of the Patterns of Religious Belief among Scientists." *Zygon* 38, no. 3 (2003) 603–32.
Brown, P. *Augustine of Hippo*. London: Faber and Faber, 1967.
Brunner, E. "Nature and Grace." In *Natural Theology: Nature and Grace and No!*, edited by J. Ballie, 15–64. London: Centenary, 1946.
Buckland, W. *Reliqiae Diluvianae*. London: Murray, 1824.
Buckley, M. J. *At the Origins of Modern Atheism*. New Haven: Yale University Press, 1987.
Burleigh, J. H. S. "The Doctrine of the Holy Spirit in the Latin Fathers." *Scottish Journal of Theology* 7, no. 2 (1954) 113–32.
Burnet, T. *The Sacred Theory of the Earth*. 1681. Reprint, London: Centaur, 1965.
Burtt, E. A. *The Metaphysical Foundations of Modern Physical Science*. London: Routledge & Kegan Paul, 1959.
Busch, E. *Karl Barth: His Life from Letters and Autobiographical Texts*. London: SCM, 1976.

Calvin, J. *Institutes of the Christian Religion*. Translated by Henry Beveridge. Grand Rapids: Eerdmans, 1995.

Cantor, G., and Hodge, M. "Introduction." In *Conceptions of Ether: Studies in the History of Ether Theories 1740–1900*, edited by G. Cantor and M. Hodge, 1–60. Cambridge: Cambridge University Press, 1981.

Cantor, G., and C. Kenny. "Barbour's Fourfold Way: Problems with His Taxonomy of Science-Religion Relationships." *Zygon* 36, no. 4 (2001) 765–81.

Cary, L. *Discourse of Infallibility*. London: Dawson, 1651.

Case-Winter, A. "The Question of God in an Age of Science: Constructions of Reality and Ultimate Reality in Theology and Science." *Zygon* 32, no. 3 (1997) 351–75.

Chadwick, O. *The Secularization of the European Mind in the Nineteenth-century*. Cambridge: Cambridge University Press, 1975.

———. *Victorian Church*. 2 vols. London: A. & C. Black, 1966.

Chambers, R. "Vestiges of the Natural History of Creation." In *Vestiges of the Natural History of Creation and Other Evolutionary Writing*, edited by J. Secord, 1–390. Chicago: University of Chicago Press, 1844.

Chapman, J. "Monophysites and Monophysitism." In *Catholic Encyclopaedia*, vol. 10, edited by K. Knight. New York: Appleton, 1911. http://www.newadvent.org/cathen/10489b.htm (accessed 14 March 2015).

Childs, B. S. *Old Testament Theology in a Canonical Context*. London: SCM, 1985.

Chillingworth, W. *Religion of Protestants a Safe Way to Salvation*. 1845 ed. London: Tegg, 1637.

Christie, J. "The Development of the Historiography of Science." In *Companion to the History of Modern Science*, edited by R. Olby, 5–22. London: Routledge, 1990.

Chung, Paul S. "Karl Barth and God in Creation: Towards an Interfaith Dialogue with Science and Religion." *Theology and Science* 3, no. 1 (2005) 55–70.

Clayton, Philip. "Biology, Directionality, and God: Getting Clear on the Stakes for Religion—Science Discussion." *Theology and Science* 4, no. 2 (2006) 121–27.

———. "The Emergence of Spirit: From Complexity to Anthropology to Theology." *Theology and Science* 4, no. 3 (2006) 291–307.

———. *The Problem of God in Modern Thought*. Grand Rapids: Eerdmans, 2000.

Clines, D. A. *The Theme of the Pentateuch*. Sheffield: JSOT, 1978.

Cohen, I. B., and R. S. Westfall. *Newton*. New York: Norton, 1995.

Coleridge, S. *The Friend and Aids to Reflection*. London: Pickering, 1848.

The Columbia Encyclopaedia. New York: Columbia University Press, 2007.

Comte, A. *The Positive Philosophy of Auguste Comte*. Translated by H. Martineau. London: Chapman, 1853.

Conway, Charles G. "Defining 'Spirit': An Encounter between Naturalists and Trans-Naturalists." *Theology and Science* 5, no. 2 (2007) 167–83.

Copernicus, N. *On the Revolutions*. Edited by E. Rosen. Baltimore: Johns Hopkins University Press, 2008.

Cortez, M. "What Does It Mean to Call Karl Barth a 'Christocentric' Theologian?" *Scottish Journal of Theology* 60, no. 1 (2007) 127–43.

Craig, W. L., and Q. Smith. *Theism, Atheism and Big Bang Cosmology* Oxford: Clarendon, 1993.

Croce, P. J. "Probabilistic Darwinism: Louis Agassiz Vs Asa Gray on Science, Religion and Certainty." *Journal of Religious History* 22 (1998) 35–58.

Cuvier, G. "Living and Fossil Elephants." In *Georges Cuvier, Fossil Bones, and Geological Catastrophes: New Translations & Interpretations of the Primary Texts*, edited by M. J. S. Rudwick, 13–24. Chicago: University of Chicago Press, 1997.
———. "Megatherium from South America." In *Georges Cuvier, Fossil Bones, and Geological Catastrophes: New Translations & Interpretations of the Primary Texts*, edited by M. J. S. Rudwick, 25–32. Chicago: University of Chicago Press, 1997.
Daley, B. E. "The Origenism of Leontius of Byzantium." *Journal of Theological Studies* 27 (1976) 333–69.
———. "A Richer Union: Leontius of Byzantium and the Relationship of Human and Divine in Christ." *Studia Patristica* 24 (1993) 239–65.
Danaher, William, J. Jr. "Intersections: Science, Theology, and Ethics." *Anglican Theological Review* 81, no. 2 (1999) 352.
Darwin, C. *The Autobiography of Charles Darwin and Selected Letters*. Edited by F. Darwin. New York: Dover, 1958.
———. *Correspondence of Charles Darwin*. 15 vols. Edited by F. Burkhardt. Cambridge: Cambridge University Press, 1985.
———. "Darwin, C. R. To McDermott, F. A., 24 Nov 1880." Darwin Correspondence Project, 2015, http://www.darwinproject.ac.uk/ (accessed May 20, 2008).
———. *The Descent of Man*, London: Murray, 1871.
———. "Letter 5307—Darwin, C. R. to Boole, M. E., 14 Dec 1866." Darwin Correspondence Project, http://www.darwinproject.ac.uk/entry-5307/ (accessed January 20, 2010).
———. *The Life and Letters of Charles Darwin*. 3 vols. Edited by F. Darwin. London: Murray, 1887.
———. *A Monograph on the Sub-Class Cirripedia: With Figures of All the Species*. London: Ray Society, 1854.
———. "The Position of the Bones of Mastodon (?) at Port St Julian is of Interest." CUL-DAR42.97-99 (Feb. 5, 1835), Darwin Correspondence Project, http://www.darwinproject.ac.uk/ (accessed October 15, 2008).
———. *Origin of Species*. London: Collins, 1859.
———. *The Variation of Plants and Animals under Domestication*. London: Murray, 1875.
Davies, P. *The Mind of God*. London: Penguin, 1992.
———. "Physics and the Mind of God: The Templeton Prize Address." *First Things* 55 (1995) 31–35.
———. *Runaway Universe*. London: Penguin, 1978.
Davis, Edward B. "Christianity and Early Modern Science: Beyond War and Peace?" *Perspectives on Science & Christian Faith*, no. 46 (1994) 133–35.
Davis, J. N., and Daly, M. "Evolutionary Theory and the Human Family." *Quarterly Review of Biology* 72, no. 4 (1997) 407–35.
Dawkins, Richard. *The God Delusion*. 2nd ed. London: Bantam, 2006.
———. "Obscurantism to the Rescue." *Quarterly Review of Biology* 72, no. 4 (1997) 397–99.
———. *The Selfish Gene*. 2nd ed. Oxford: Oxford University Press, 2006.
De Smet, R., and K. Verelst. "Newton's Scholium Generale: The Platonic and Stoic Legacy—Philo, Justus Lipsius and the Cambridge Platonists." *History of Science* 39 (2001) 1–30.

Dempsey, L. "Written in the Flesh: Isaac Newton on the Mind-Body Relation." *Studies in History and Philosophy of Science* 37 (2006) 420–41.
Descartes, R. "Descartes to Mersenne, 15 April 1630." In *The Philosophical Writings of Descartes*, edited by J. Cottingham, 3:23. Cambridge: Cambridge University Press, 1984.
Desmond, A. *Huxley: The Devil's Disciple*. London Joseph, 1994.
———. *Huxley: Evolution's High Priest*. London: Joseph, 1997.
Desmond, A., and J. Moore. *Darwin*. London: Joseph, 1991.
Dobbs, B. J. "Newton's Alchemy and His Theory of Matter." *Isis* 73 (1982) 512–28.
Dodds, Michael J. "Hylomorphism and Human Wholeness: Perspectives on the Mind-Brain Problem." *Theology and Science* 7, no. 2 (2009) 141–62.
Draper, J. W. *History of the Conflict between Religion and Science*. London: Paul, 1882.
Eco, U. *The Name of the Rose*. London: Minerva, 1983.
Edwards, D. "Christology in the Meeting between Science and Religion. A Tribute to Ian Barbour." *Theology and Science* 3, no. 2 (2005) 211–20.
———. "Evolution and the Christian God." In *Interdisciplinary Perspectives on Cosmology and Biological Evolution*, edited by L. A. Trost, 174–94, Adelaide: Centre for Theology and the Natural Sciences, Berkley, 2001.
Fara, P. *Newton: The Making of Genius*. London: Macmillan, 2002.
Fauvel, J., Flood, R., Shortland, M. and Wilson, R. *Let Newton Be!* Oxford: Oxford University Press, 1988.
Florovsky, G. *The Byzantine Fathers of the Fifth-Century*. Vaduz: Buchervertriebsansalt, 1987.
Force, J. E. "The God of Abraham and Isaac." In *The Books of Nature and Scripture: Recent Essays on Natural Philosophy, Theology and Biblical Criticism in the Netherlands of Spinoza's Time and the British Isles of Newton's Time*, edited by Force, J. E. and Popkin, R. H. 179–200. London: Kluwer Academic, 1994.
———. "Natural Law, Miracles and Newtonian Science." In *Newton and Newtonianism: New Studies*, edited by J. E. Force and S. Hutton, 65–92. Dordrecht: Kluwer, 2004.
Force, J. E., and S. Hutton, eds. *Newton and Newtonianism: New Studies*. Dordrecht: Kluwer, 2004
Force, J. E., and R. H. Popkin, eds. *The Books of Nature and Scripture: Recent Essays on Natural Philosophy, Theology and Biblical Criticism in the Netherlands of Spinoza's Time and the British Isles of Newton's Time*. London: Kluwer Academic, 1994.
Foster, M. B. "The Christian Doctrine of Creation and the Rise of Modern Science." *Mind* 43, no. 172 (1934) 446–68.
———. "Christian Theology and Modern Science of Nature (I.)." *Mind* 44, no. 176 (1934) 439–66.
———. "Christian Theology and Modern Science of Nature (II.)." *Mind* 45, no. 177 (1934) 1–27.
Fredouille, J. C. *Tertullien et la Conversion de la Culture Antique*. Paris: Études Augustiniennes, 1972.
Frei, H. W. *The Eclipse of Biblical Narrative: A Study in Eighteenth and Nineteenth-century Hermeneutics*. New Haven: Yale University Press, 1974.
Frend, W. H. C. *The Rise of Christianity*. Philadelphia: Fortress, 1984.
Gaussen, L. *Theopneustia: The Plenary Inspiration of the Holy Scriptures*. Chicago: Bible Institute Colportage Assn., 1841.
Gillespie, C. G. *Genesis and Geology*. New York: Harper and Bros., 1951.

Gould, S. J. "Ontogeny and Phylogeny—Revisited and Reunited." *BioEssays* 14, no. 4 (1992) 275–79.
Grabbe, L. L. *Priest, Prophets, Diviners, Sages: A Socio-Historical Study of Religious Specialists in Ancient Israel*. Valley Forge, PA: Trinity, 1995.
Gray, A. *Natural Selection Not Inconsistent with Natural Theology*. London: Trubner, 1861.
———. "The Origin of Species by Means of Natural Selection." Darwin Correspondence Project, University of Cambridge, 2015. http://www.darwinproject.ac.uk/gray-review-the-origin-of-species (accessed Nov 1, 2014).
Greaves, R. L. "Puritanism and Science: Anatomy of a Controversy." *Journal of the History of Ideas* 30 (1969) 345–68.
Green, J. "Restoring the Human Person: New Testament Voices for a Wholistic and Social Anthropology." In *Neuroscience and the Person: Scientific Perspectives on Divine Action*, edited by R. J. Russell et al., 3–22. Vatican: Vatican Observatory and the Center for Theology and the Natural Sciences, 1999.
Gregory of Nazianzus, "Epistle 101." In *Christology of the Later Fathers*, edited by E. Hardy, 215–24. London: SCM, 1954.
Grenz, S. J., and Olson, R. E. *20th Century Theology*. Downers Grove IL: InterVarsity, 1992.
Grillmeier, A. "The Christology of Leontius of Byzantium: His Contribution to Solving the Chalcedonian Problem." In *Christ in Christian Tradition*, edited by A. Grillmeier, 2:185–312. London: Mowbray, 1995.
Gunton, C. *Christ and Creation*. Carlisle, UK: Paternoster, 1993.
Hall, R., and M. Boas Hall. "Newton and the Theory of Matter." In *Newton*, edited by I. B. Cohen and R. S. Westfall, 72–87. New York: Norton, 1983.
Harrison, P. *The Bible, Protestantism and the Rise of Natural Science*. Cambridge: Cambridge University Press, 1998.
———. "The Book of Nature and Early Modern Science." In *The Book of Nature in Early Modern and Modern History*, edited by K. van Berkel, and V. Vanderjagt, 1–26, Leuven: Peeters, 2006.
———. "Curiosity, Forbidden Knowledge, and the Reformation of Natural Philosophy in Early Modern England." *Isis* 95, no. 2 (2001) 265–88.
———. *The Fall of Man and the Foundation of Science*. Cambridge: Cambridge University Press, 2007.
———. "Newtonian Science, Miracles, and the Laws of Nature." *Journal of the History of Ideas* 56 (1998) 531–53.
———. "Religion, the Royal Society, and the Rise of Science." *Theology and Science* 6, no. 3 (2008) 255–57.
———. "'Science and Religion': The Constructing the Boundaries." *The Journal of Religion* 86, no. 1 (2006) 81–107.
———. "Voluntarism and Early Modern Science." *History of Science* 40 no. (2002) 63–89
———. "Was Newton a Voluntarist?" In *Newton and Newtonianism: New Studies*, edited by J. E. Force, and S. Hutton, 39–64. Dordrecht: Kluwer, 2004.
Haught, J. F. "In Praise of Imperfection." *Theology and Science* 6, no. 2 (2008) 173–77.
Henry, J. "'Pray Do Not Ascribe That Notion to Me': God and Newton's Gravity." In *The Books of Nature and Scripture: Recent Essays on Natural Philosophy, Theology and Biblical Criticism in the Netherlands of Spinoza's Time and the British Isles of*

Newton's Time, edited by J. E. Force, and R. H. Popkin, 123–49. London: Kluwer Academic, 1994.

———. *The Scientific Revolution and the Origins of Modern Science*. Basingstoke, UK: Palgrave, 2002.

Hewlett, Martinez, and Peters, T. "Why Darwin's Theory of Evolution Deserves Theological Support." *Theology and Science* 4, no. 2 (2006) 171–82.

The Hobart Town Magazine. Vol. 3. Hobart Town, van Dieman's Land: Melville, 1834.

Hooykaas, R. *Religion and the Rise of Modern Science*. Edinburgh: Scottish Academic, 1972.

Hume, D. *An Enquiry Concerning the Human Understanding*, London: Millar, 1748.

Hunsinger, G. *Disruptive Grace: Studies in the Theology of Karl Barth*. Grand Rapids: Eerdmans, 2000.

———. *How to Read Karl Barth*. New York: Oxford University Press, 1991.

———. "Truth as Self-Involving: Barth and Lindbeck on the Cognitive and Performative Aspects of Truth in Theological Discourse." *Journal of the American Academy of Religion* 61, no. 1 (1993) 41–56.

Hutton, C. *A Philosophical and Mathematical Dictionary*. Vol. 2. London: F. C. & J. Rivington, 1815.

Hutton, S. "More, Newton and the Language of Biblical Prophecy." In *The Books of Nature and Scripture: Recent Essays on Natural Philosophy, Theology and Biblical Criticism in the Netherlands of Spinoza's Time and the British Isles of Newton's Time*, edited by J. E. Force and R. H. Popkin, 39–53. London: Kluwer Academic, 1994.

Huxley, T. H. "Agnosticism and Christianity." Edited by C. Blinderman and D. Joyce. *Collected Essays*, 1899. The Huxley File. http://aleph0.clarku.edu/huxley/CE5/Agn-X.html (accessed March 15, 2015).

———. "An Apologetic Irenicon." Edited by C. Blinderman and D. Joyce. *Fortnightly Review*, November 1892. The Huxley File. http://aleph0.clarku.edu/huxley/UnColl/Rdetc/IREN.html (accessed March 15, 2015).

———. "The Bible and Modern Criticism." Edited by C. Blinderman and D. Joyce. *The Times*, January-February 1892. The Huxley File. http://aleph0.clarku.edu/huxley/UnColl/LonTimes/Bib-MCr.html (accessed March 15, 2015).

———. "Bishop Berkley on the Metaphysics of Sensation." Edited by C. Blinderman and D. Joyce. *Collected Essays*, 1871. The Huxley File. http://aleph0.clarku.edu/huxley/CE6/Berk.html (accessed March 15, 2015).

———. "The Evolution of Theology." Edited by C. Blinderman and D. Joyce. *Collected Essays*, 1886. The Huxley File. http://aleph0.clarku.edu/huxley/CE4/EvTheo.html (accessed March 15, 2015).

———. "Has a Frog a Soul?" Edited by C. Blinderman and D. Joyce. *Metaphysical Society (8 November 1870)*. The Huxley File. http://aleph0.clarku.edu/huxley/Mss/FROG.html (accessed March 15, 2015).

———. "Lectures on Evolution." Edited by C. Blinderman and D. Joyce. *Collected Essays*, 1877. The Huxley File. http://aleph0.clarku.edu/huxley/CE4/LecEvol.html (accessed March 15, 2015).

———. "On the Present State of Knowledge as the Structure and Functions of Nerve." Edited by C. Blinderman and D. Joyce. *Scientific Memoirs*, 1854. The Huxley File. http://aleph0.clarku.edu/huxley/SM1/Nerve.html (accessed March 15, 2015).

———. "On Sensation and the Unity of Structure of Sensiferous Organs." Edited by C. Blinderman and D. Joyce. *Collected Essays*, 1879. The Huxley File. http://aleph0.clarku.edu/huxley/CE6/Sense.html (accessed March 15, 2015).

———. "Vestiges of the Natural History of Creation." Edited by C. Blinderman and D. Joyce. *Scientific Memoirs*, 1854. The Huxley File. http://aleph0.clarku.edu/huxley/SM5/vest.html (accessed March 15, 2015).

———. "Witness to the Miraculous." Edited by C. Blinderman and D. Joyce. *Collected Essays*, 1886. The Huxley File. http://aleph0.clarku.edu/huxley/CE5/Wit.html (accessed March 15, 2015).

Iliffe, R. "Digitizing Isaac: The Newton Project and Electronic Edition of Newton's Papers." In *Newton and Newtonianism: New Studies*, edited by J. E. Force and S. Hutton, 23–38. Dordrecht: Kluwer, 2004.

———. *Newton: A Very Short Introduction*. Oxford: Oxford University Press, 2007.

———. "Prosecuting Athanasius: Protestant Forensics and the Mirrors of Persecution." In *Newton and Newtonianism: New Studies*, edited by J. E. Force and S. Hutton, 113–54. Dordrecht: Kluwer, 2004.

Irenaeus. *Against Heresies*. In *Ante-Nicene Fathers*, edited by A. Roberts and J. Donaldson, 1:640–1175. Edinburgh: T. & T. Clark, 1993.

Jackelén, Antje. "What Theology Can Do for Science." *Theology and Science* 6, no. 3 (2008) 287–303.

Jacob, M. "Introduction." In *Newton and Newtonianism: New Studies*, edited by J. E. Force and S. Hutton, ix–xvii. Dordrecht: Kluwer, 2004.

Jaki, S. *Means to Message: A Treatise on Truth*. Grand Rapids: Eerdmans, 1999.

Jastrow, R. *God and the Astronomers* New York: Norton, 1978.

Jeannerod, M. "Are there Limits to the Naturalization of Mental States?" In *Neuroscience and the Person: Scientific Perspectives on Divine Action*, edited by R. J. Russell et al., 119–27. Vatican: Vatican Observatory and the Center for Theology and the Natural Sciences, 1999.

Jeans, J. *The Growth of the Physical Sciences*. London: Readers Union, 1950.

Jeffreys, Derek. "A Counter-Response to Nancey Murphy on Non-Reductive Physicalism." *Theology and Science* 3, no. 1 (2005) 94–87.

———. "The Soul Is Alive and Well: Nonreductive Physicalism and Emergent Mental Properties." *Theology and Science* 2, no. 2 (2004) 205–25.

John Paul II. "Le Message a L'Académie Pontificale Des Sciences." *Quarterly Review of Biology* 72, no. 4 (1997) 377–80.

———. "Message to the Pontifical Academy of Sciences." *Quarterly Review of Biology* 72, no. 4 (1997) 381–85.

Kant, I. *The Critique of Pure Reason*. London: Longmans, Green, 1909.

Kelly, J. N. D. *Jerome*. London: Duckworth, 1975.

———. *Early Christian Doctrines*. London: A. & C. Black, 1977.

Klinefelter, Donald, S. "E. O. Wilson and the Limits of Ethical Naturalism." *American Journal of Theology & Philosophy* 21, no. 3 (2000) 240–55.

Knox, K. C. "Dephlogisticating the Bible: Natural Philosophy and Religious Controversy in Late Georgian Cambridge." *History of Science* 34 (1996) 167–200.

Koyre, A. "The Significance of the Newtonian Synthesis." In *Newtonian Studies*, edited by A. Koyre, 6–7. Cambridge, MA: Harvard University Press, 1965.

Kraeling, E. *The Old Testament Since the Reformation*. London: Lutterworth, 1955.

Kragh, H. *An Introduction to the Historiography of Science*. Cambridge: Cambridge University Press, 1987.

Kuczewski, Mark. "Two Models of Ethical Consensus, or What Good Is a Bunch of Bioethicists?" *Cambridge Quarterly of Healthcare Ethics* 11, no. 1 (2002) 27.

Laats, A. *Doctrines of the Trinity in Eastern and Western Theologies: A Study with Special Reference to K. Barth and V. Lossky*. New York: Lang, 1999.

Lakatos, I. "History of Science and Its Rational Reconstructions." In *The Methodology of Scientific Research Programmes*, edited by I. Lakatos, I. J. Worrall, and G. Currie, 102–38. Cambridge: Cambridge University Press, 1978.

Lambert, Kevin. "Fuller's Folly, Kuhnian Paradigms, and Intelligent Design." *Social Studies of Science* 36, no. 6 (2006) 835–42.

Lang, U. M. "*Anhypostasis-Enhypostasis*: Church Fathers, Protestant Orthodoxy and Karl Barth." *Journal of Theological Studies* 49, no. 2 (1998) 630–58.

Larson, E. J., and L. Witham. "Scientists Are Still Keeping the Faith." *Nature* 386, no. April (1996) 433–37.

Lash, N. *The Beginning and the End of Religion*. Cambridge: Cambridge University Press, 1996.

———. *Believing Three Ways in One God: A Reading of the Apostles Creed*. Notre Dame, IN: University of Notre Dame Press, 1993.

———. "Theory Theology and Ideology." In *The Sciences and Theology in the Twentieth-century*, edited by A. R. Peacocke, 209–28. Notre Dame, IN: University of Notre Dame Press, 1981.

Latourette, K. S. *The Nineteenth-century in Europe*. Vol. 2. Grand Rapids: Zondervan, 1959.

LeDoux, J. "Emotions—A View Through the Brain." In *Neuroscience and the Person: Scientific Perspectives on Divine Action*, edited by R. J. Russell et al., 101–18. Vatican: Vatican Observatory and the Center for Theology and the Natural Sciences, 1999

———. "Emotions: How I've Looked for Them in the Brain," In *Neuroscience and the Person: Scientific Perspectives on Divine Action*, edited by R. J. Russell et al., 41–44. Vatican: Vatican Observatory and the Center for Theology and the Natural Sciences, 1999.

Leibniz, G. W. "Discourse on Metaphysics." In *Discourse on Metaphysics and the Monadology*, edited by G. R. Montgomery, 3–63. Buffalo: Prometheus, 1992.

———. *Discourse on Metaphysics and the Monadology*. Buffalo: Prometheus, 1992.

———. "Monadology." In *Discourse on Metaphysics and the Monadology*, edited by G. R. Montgomery, 67–88. Buffalo: Prometheus, 1992.

Lennox, J. C. *God's Undertaker: Has Science Buried God?* Oxford: Lion, 2007.

Leontius. "*Liber Tres Contra Nestorianos Et Eutychianos.*" In *Patrologia Graeca*, edited by J.-P. Migne, 86a:1267–395. Paris: Garnier, 1865.

———. "*Adv. Severum.*" In *Patrologia Graeca*, edited by Migne, J.-P. 86b:1944. Paris: Garnier, 1865.

LeRon Shults, F "Anglo-American Postmodernity: Philosophical Perspectives on Science, Religion, and Ethics." *Theology Today* 55, no. 2 (1998) 255–56.

——— "A Dubious Christological Formula: From Leontius of Byzantium to Karl Barth." *Zygon* 57, no. 4 (1996) 431–36.

Lightman, B. *The Origins of Agnosticism: Victorian Unbelief and the Limits of Knowledge*. Baltimore: Johns Hopkins University Press, 1987.

Lindbeck, G. A. *The Nature of Doctrine: Religion and Theology in a Postliberal Age.* London: SPCK, 1984.
Lindberg, D. C. *The Beginnings of Western Science.* Chicago: University of Chicago Press, 1992.
Lodge, O. ed. *Pioneers of Science.* 1926 London: Macmillan, 1893.
Lovett, R. *Philosophical Essays.* Worcester: Lewis, 1766.
Lyell, Sir C. *Principles of Geology.* 11th ed. New York: Appleton, 1883.
MacIntyre, A. *After Virtue.* Notre Dame: Notre Dame University Press, 1981.
———. "The Fate of Theism." In *The Religious Significance of Atheism*, edited by A. C. MacIntyre and P. Ricoeur, 1–45 New York: Columbia University Press, 1969.
Maclaurin, C. *An Account of Newton's Philosophical Discoveries.* London: Nourse et al., 1775.
Mandelbrote, S. "A Duty of the Greatest Moment: Isaac Newton and the Writing of Biblical Criticism." *British Journal for the History of Science* 26 (1993) 281–302.
———. "Eighteenth-Century Reaction's to Newton's Anti-Trinitarianism." In *Newton and Newtonianism: New Studies*, edited by J. E. Force and S. Hutton, 93–112. Dordrecht: Kluwer, 2004.
Manuel, F. E. *The Religion of Isaac Newton.* Oxford: Clarendon, 1974.
Marcum, James. "Exploring the Rational Boundaries between the Natural Sciences and Christian Theology 1." *Theology and Science* 1, no. 2 (2003) 203–20.
Marshall, I. H. *Biblical Inspiration.* London: Hodder and Stoughton, 1982.
Martin, B. *A New and Comprehensive System of the Newtonian Philosophy, Astronomy and Geography.* London: Micklewright, 1746.
Masters, R. D. and Paul M. Churchland. "Neuroscience and Human Nature the Engine of Reason, the Seat of the Soul: A Philosophical Journey into the Brain." *Quarterly Review of Biology* 72, no. 4 (1997) 448–50.
McCaul, A. *Testimonies to the Divine Authority and Inspiration of the Holy Scriptures.* London: Rivingtons, 1862.
McCormack, B. L. *Karl Barth's Critically Realistic Dialectical Theology.* Oxford: Oxford University Press, 1995.
McDonnell, K. "Communion Ecclesiology and Baptism in the Spirit: Tertullian and the Early Church." *Theological Studies* 49 (1988) 671–93.
McEvoy, J. "Positivism, Whiggism, and the Chemical Revolution: A Study in the Historiography of Chemistry." *History of Science* 35 (1997) 1–33.
McGrath, A. *Christian Theology an Introduction.* Oxford: Blackwell, 1994.
———. *Scientific Theology.* Vol. 2, *Reality.* Grand Rapids: Eerdmans, 2002.
———. *The Twilight of Atheism.* New York: Doubleday, 2004.
McGuire, J. E., and M. Tamny. *Certain Philosophical Questions: Newton's Trinity Notebook.* Cambridge: Cambridge University Press, 1983.
McGuire, J. E., and P. M. Rattansi. "Newton and the 'Pipes of Pan.'" In *Newton*, edited by I. B. Cohen and R. S. Westfall, 96–108. New York: Norton, 1995.
McIntosh, A. "The Doctrine of Appropriation as an Interpretative Framework for Karl Barth's Pneumatology of the Church Dogmatics." *Pacifica* 20, no. 3 (2007) 278–90.
McLachlan, H. *Sir Isaac Newton: Theological Manuscripts.* Liverpool: University Press, 1950.
Merton, R. *Science, Technology and Society in the Seventeenth-century England.* New York: Harper & Row, 1970.
Metzger, B. M. *The Canon of the New Testament.* Oxford: Clarendon, 1987.

Moore, J. R. *The Post Darwinian Controversies*. Cambridge: Cambridge University Press, 1979.

Moritz, Joshua M. "Science and Religion: A Fundamental Face-Off, or Is There a Tertium Quid?" *Theology and Science* 6, no. 2 (2008) 137–45.

Morton, P. "Darwinism and the Victorian Literary Imagination: A Bibliography." Architecture of Modern Political Power, Nov. 8, 2012, http://www.mega.nu:8080/ampp/PeterMorton/darwin_biblio.htm (accessed 23 May 2008).

———. *The Vital Science: Biology and the Literary Imagination 1860–1900*. London: Allen & Unwin, 1984.

Murphy, G. L. "Science as Goad and Guide for Theology." *Dialog* 46, no. 3 (2007) 225–34

Murphy, N. Darwin, Social Theory, and the Sociology of Scientific Knowledge." *Zygon* 34, no. 4 (1999) 573–642.

———. "How Physicalists Avoid Being Reductionists." In *Interdisciplinary Perspectives on Cosmology and Biological Evolution*, edited by L. A. Trost. 69–90 Adelaide: Centre for Theology and the Natural Sciences, Berkley, 2001.

———. "On the Role of Philosophy in Theology-Science Dialogue." *Theology and Science* 1, no. 1 (2003) 79–93.

———. "Response to Derek Jeffreys." *Theology and Science* 2, no. 2 (2004) 227–30.

———. "What Has Theology to Learn from Scientific Methodology?" In *Science and Theology: Questions at the Interface*, edited by Rae, M. Regan, H. and Stenhouse, J. 101–37. Edinburgh: T. & T. Clark, 1994.

———. "Why Christians Should Be Physicalists." In *Interdisciplinary Perspectives on Cosmology and Biological Evolution*, edited by L. A. Trost, 52–68 Adelaide: Centre for Theology and the Natural Sciences, Berkley, 2001.

Nebelsick, H. P. "Karl Barth's Understanding of Science." In *Theology Beyond Christendom*, edited by J. Thompson, 165–214. Allison Park, PA: Pickwick, 1986.

———. *Theology and Science in Mutual Modification*. New York: Oxford University Press, 1980.

Needham, J. *Science, Religion, and Socialism* Christ and the Social Revolution, edited by J. Lewis. London: Left Book Club, 1937.

———. *Science Religion and Reality*. Port Washington, NY: Kennikat, 1970.

Newbigin, L. *The Gospel in a Pluralist Society*. Grand Rapids: Eerdmans, 1987.

Newton, I. *The Correspondence of Isaac Newton*. Edited by H. W. Turnbull. 7 vols. Cambridge: Cambridge University Press, 1959–1978.

———. *De Aere et Aethere*. In *Newton*, edited by I. B. Cohen and R. S. Westfall, 34–39. New York: Norton, 1679.

———. *De Gravitatione*. In *Isaac Newton: Philosophical Writings*, edited by A. Janiak, 12–39. Cambridge: Cambridge University Press, 2004.

———. "Drafts on the History of the Church" (Section 7). Yahuda ms. 15.7, National Library of Israel, Jerusalem. The Newton Project, April 2007. http://www.newtonproject.sussex.ac.uk/view/texts/normalized/THEM00237 (accessed March 15, 2015).

———. "The Language of the Prophets." In *Sir Isaac Newton: Theological Manuscripts*, edited by H. McLachlan, 119–26. Liverpool: University Press, 1950.

———. "Of Colours." Add. ms. 3975, Cambridge University Library. The Newton Project, October 25, 2006. http://www.newtonproject.sussex.ac.uk/texts/viewtext.php?id=NATP00048 (accessed July 8, 2008).

---. "Of Natures Obvious Laws & Processes in Vegetation." Edited by William R. Newman. Dibner MS. 1031 B SCDIRB. The Chymistry of Isaac Newton, 2006. http://purl.dlib.indiana.edu/iudl/newton/ALCH00081 (accessed 14 March 2015).

---. "Paradoxical Questions Concerning the Morals and Action of Athanasius and His Followers." In *Sir Isaac Newton: Theological Manuscripts*, edited by H. McLachlan, 60–118. Liverpool: University Press, 1950.

---. *The Principia*. New York: Prometheus, 1995.

---. *The Prophecies of Daniel and the Apocalypse*. Hyderabad, India: Printland, 1998.

---. "Queries Regarding the Word '*Homooousios*.'" In *Sir Isaac Newton: Theological Manuscripts*, edited by H. McLachlan, 44–47. Liverpool: University Press, 1950.

---. "Questions from the Optics." In *Newton's Philosophy of Nature: Selections from His Writings*, edited by H. S. Thayer, 135–79. New York: Macmillan, 1953.

---. "Questiones Quædam Philosophiæ." Add. ms. 3996, Cambridge University Library. The Newton Project, October 4–8, 2007. http://www.newtonproject.sussex.ac.uk/texts/viewtext.php?id=THEM00092 (accessed June 10, 2008).

---. "Theological Notebook." Keynes Ms. 2., Cambridge University Library. The Newton Project, 2003. http://www.newtonproject.sussex.ac.uk/texts/viewtext.php?id=THEM00180 (accessed June 10, 2008).

---. *Two Notable Corruptions of Scripture* London: Green, 1841.

Noble, S. *The Plenary Inspiration of the Scriptures*. London: Simpkin and Marshall, 1825.

Noakes, R. "Recreating Newton: Newton Biography and the Making of Nineteenth-century Science." *Victorian Studies* 51, no. 1 (2008) 168–71.

Nock, A. D. "Tertullian and the Ahori." *Viligiae Christianae* 4, no. 1 (1950) 129–41

Noll, M. A. and D. N. Livingstone. *B B Warfield: Evolution, Science and Scripture, Selected Writings*. Grand Rapids: Eerdmans, 2000.

Norton, A. *The Genuineness of the Gospels*. London: Chapman, 1847.

Numbers, R. L. *The Creationists: From Scientific Creationism to Intelligent Design*. Cambridge, MA: Harvard University Press, 1986.

---. *The Creationists: From Scientific Creationism to Intelligent Design*. Rev. ed. Cambridge, MA: Harvard University Press, 2006.

Numbers, R. L., and D. C. Lindberg. *God and Nature: Historical Essays on the Encounter between Christianity and Science*. Berkley: University of California Press, 1986.

Oakley, Francis "Christian Theology and the Newtonian Science: The Rise of the Concept of the Laws of Nature." *Church History* 30, no. 4 (1961) 433–57.

Osler, M. "Mixing Metaphors: Science and Religion or Natural Philosophy and Theology in Early Modern Europe." *History of Science* 35 (1997) 91–113.

---. "The New Newtonian Scholarship and the Fate of the Scientific Revolution." In *Newton and Newtonianism: New Studies*, edited by J. E. Force and S. Hutton, 1–14. Dordrecht: Kluwer, 2004.

Origen. "*De Principis*." In *Ante-Nicene Fathers*, edited by A. Roberts and J. Donaldson, 4:497–780. Edinburgh: T. & T. Clark, 1994.

Osborn, E. *Tertullian, First Theologian of the West*. Cambridge: Cambridge University Press, 1997.

Ospovat, D. *The Development of Darwin's Theory*, Cambridge: Cambridge University Press, 1981.

Owen, M. *Witness of Faith*. Melbourne: Uniting Church, 1984.

Paley, W. *Natural Theology*, New York: Turner, Hughes and Hayden, 1843.

———. *A View of the Evidences of Christianity*. Edited by T. R. Birks. London: London Tract Society, 1859.
Pannenberg, W. "God as Spirit—and Natural Science." *Zygon* 36, no. 4 (2001) 783–94.
———. *Revelation as History*. London: Sheed and Ward, 1979.
———. *Toward a Theology of Nature*. Louisville: Westminster John Knox, 1993.
Passmore, J. *The Perfectibility of Man*. Indianapolis: Liberty Fund, 2000.
Peacocke, A. "Science and the Future of Theology: Critical Issues." *Zygon* 35, no. 14 (2000) 119–40.
———. *The Sciences and Theology in the Twentieth-century*. Notre Dame, IN: University of Notre Dame Press, 1981.
Peckhaus, V. "Logic and Metaphysics: Heinrich Scholz and the Scientific World-view," *Philosophia Mathematica* 16, no. 1 (2008) 78–99.
Pellegrino, E. D. "Theology and Evolution in Dialogue." *Quarterly Review of Biology* 72, no. 4 (1997) 385–89.
Peterson, Gregory R. "In Praise of Folly? Theology and the University." *Zygon* 43, no. 3 (2008) 563–77.
Pfizenmaier, T. C. "Was Isaac Newton an Arian?" *Journal of the History of Ideas* 58, no. (1997) 57–80.
Plantinga, A, *Where the Conflict Really Lies: Science, Religion and Naturalism*, Oxford: Oxford University Press, 2011.
Polkinghorne, J. *"Beyond Science" the Wider Human Context*. Cambridge: Cambridge University Press, 1998.
———. "Physics and Metaphysics in a Trinitarian Perspective." *Theology and Science* 1, no. 1 (2003) 33–49.
———. *Quantum Physics and Theology: An Unexpected Kinship*. London: Yale University Press, 2007.
———. *The Quantum World*. London: Longman, 1984.
———. *Science and Christian Belief*. London: SPCK, 1994.
———."Theological Notions of Creation and Divine Causality." In *Science and Theology: Questions at the Interface*, edited by M. Rae, H. Regan, and J. Stenhouse, 225–37. Edinburgh: T. & T. Clark, 1994.
Popkin, R. "Plans for Publishing Newton's Religious and Alchemical Manuscripts, 1982–1998." In *Newton and Newtonianism: New Studies*, edited by J. E. Force and S. Hutton, 15–22. Dordrecht: Kluwer, 2004.
Priestley, J. *Disquisitions Relating to Matter and Spirit* Vol. 2. Birmingham, UK: Pearson and Rollaston, 1782.
Principe, L. "Reflections on Newton's Alchemy in Light of the New Historiography of Alchemy." In *Newton and Newtonianism: New Studies*, edited by J. E. Force and S. Hutton, 205–20. Dordrecht: Kluwer, 2004.
"Proem." *Quarterly Review of Biology* 72, no. 4 (1997) 376–77.
Puddlefoot, J. "The Relationship of Natural Order to Divine Truth and Will." In *Science and Theology: Questions at the Interface*, edited by M. Rae, H. Regan, and J. Stenhouse, 148–79. Edinburgh: T. & T. Clark, 1994.
Rae, M., H Regan, and J. Stenhouse, eds. *Science and Theology: Questions at the Interface*. Edinburgh: T. & T. Clark, 1994.
Raleigh, Sir W. *The History of the World*. London: Macmillan, 1614.
Rankin, D. *Tertullian and the Church*. Cambridge: Cambridge University Press, 1995.

Raschko, Michael "Anticipation in Spirit and Nature: John Haught's Use of the Ontological Argument." *Theology and Science* 6, no. 3 (2008) 331–39

Rees,S., "Leontius of Byzantium and His Defence of the Council of Chalcedon." *Harvard Theological Review*, 24, no. 2 (1931) 111–19.

Reid, G., "Biblical Criticism (Higher)." In *Catholic Encyclopaedia*, vol. 4, edited by K. Knight. New York: Appleton, 1908. http://www.newadvent.org/cathen/04491c.htm (accessed March 14, 2015).

Rendtorff, R. *Canon and Theology: Overtures to an Old Testament Theology*. Minneapolis: Fortress, 1993.

Reventlow, H. G. *The Authority of the Bible and the Rise of the Modern World*. London SCM, 1984.

Robinson, J. M. *The Beginnings of Dialectic Theology*. Richmond: Knox, 1968.

Rogers, K. *Perfect Being Theology*. Edinburgh: Edinburgh University Press, 2002.

Rosato, P. J. *The Spirit as Lord: The Pneumatology of Karl Barth*. Edinburgh: T. & T. Clark, 1981.

Ross, H. *The Creator and the Cosmos*. Colorado Springs: Navpress, 1993.

Runia, K. *Karl Barth's Doctrine of Holy Scripture*. Grand Rapids: Eerdmanns, 1962.

Rupke, N. A. *The Great Chain of History*. Oxford: Clarendon, 1983.

Ruse, Michael. "An Evolutionist Thinks About Religion." *Theology and Science* 6, no. 2 (2008) 165–71.

———. "John Paul II and Evolution." *Quarterly Review of Biology* 72, no. 4 (1997) 391–400.

Russell, R. J., et al., eds. *Neuroscience and the Person: Scientific Perspectives on Divine Action*. Vatican: Vatican Observatory and the Center for Theology and the Natural Sciences, 1999.

Sanday, W. *Inspiration*. The 1893 Brampton Lectures. London: Longman, Green, 1903.

Santayana, G. *Reason in Common Sense*. Vol. 1, *The Life of Reason*. 1905. Reprint, New York: Dover, 1980.

Sasse, H. "Concerning the Nature of Inspiration." *Reformed Theological Review Australia* 23, no. 2 (1964) 33–43.

———. "Inspiration and Inerrancy." *Reformed Theological Review Australia* 19, no. 2 (1960) 33–48.

———. "The Rise of the Dogma of Holy Scripture." *Reformed Theological Review Australia* 18, no. 2 (1959) 44–54

Sasse, H., J. Horne, and C. Dixon. *Cassell's New Compact German-English English-German Dictionary*. Sydney: Cassell, 1966.

P. Schaff. *The Creeds of Christendom, with a History and Critical Notes*. Vol. 1. 6th ed. Grand Rapids: Harper Bros., 1877.

———. *History of the Christian Church*. Vol. 2. 5th ed. Grand Rapids: Eerdmans, 1981.

Schultz, J., ed. *The Lure of Fundamentalism*. Griffith Review. Meadowbrook, QLD: Griffith University, 2005.

Scott, E. C. "Creationists and the Pope's Statement." *Quarterly Review of Biology* 72, no. 4 (1997) 401–5.

Secord J. Ed. *Vestiges of the Natural History of Creation*. London: University of Chicago Press, 1994.

Shanks, N., and R. Dawkins. *God, the Devil, and Darwin: A Critique of Intelligent Design Theory*. Oxford: Oxford University Press, 2004.

Simmons, Ernest L. "Quantum Perichoresis: Quantum Field Theory and the Trinity." *Theology and Science* 4, no. 2 (2006) 137–50.
Smail, T. A. "The Doctrine of the Holy Spirit." In *Theology beyond Christendom*, edited by J. Thompson, 87–109. Allison Park, PA: Pickwick, 1986.
Smart, N. *Secular Education and the Logic of Religion*. London: Faber, 1968.
Smedes, Taede A. "Social and Ideological Roots of 'Science and Religion': A Social-Historical Exploration of a Recent Phenomenon." *Theology and Science* 5, no. 2 (2007) 185–201.
Smith, C. "From Design to Dissolution: Thomas Chalmers Debt to John Robison." *British Journal for the History of Science* 12, no. 40 (1979) 59–70.
Snobelen, S. D. "Isaac Newton, Heretic: The Strategies of a Nicodemite." *British Journal for the History of Science* 21 (1998) 381–419.
Soranus. *Soranus' Gynaecology*. Translated by O. Temkin. Baltimore: Johns Hopkins Press, 1956.
Spicer, E. E. *Aristotle's Conception of the Soul*. London: University of London Press, 1934.
Spezio, Michael L. "Interiority and Purpose: Emerging Points of Contact for Theology and the Neurosciences." *Theology and Science* 7, no. 2 (2009) 119–21.
Stein, Ross L. "The Action of God in the World—a Synthesis of Process Thought in Science and Theology." *Theology and Science* 4, no. 1 (2006) 51–69.
Stewart, L. "Seeing through the Scholium: Religion and Reading Newton in the Eighteenth-century." *History of Science* 34 (1996) 123–165.
———. "The Trouble with Newton in the Eighteenth-century." In *Newton and Newtonianism: New Studies*, edited by J. E. Force and S. Hutton, 221–38. Dordrecht: Kluwer, 2004.
Stillingfleet, E. *Origines Sacrae or a Rational Account of the Grounds of Natural and Revealed Religion*. Oxford: Mortlock, 1662.
Stoeger, W. "Cosmology and a Theology of Creation." In *Interdisciplinary Perspectives on Cosmology and Biological Evolution*, edited by L. A. Trost, 128–45, Adelaide: Centre for Theology and the Natural Sciences, Berkley, 2001.
———. "Science the Laws of Nature and Divine Action." In *Interdisciplinary Perspectives on Cosmology and Biological Evolution*, edited by L. A. Trost, 146–52, Adelaide: Centre for Theology and the Natural Sciences, Berkley, 2001.
Teilhard de Chardin, P. *The Phenomenon of Man*. London: Harper & Row, 1959.
Tertullian. *Adversus Marcion*. In *Ante-Nicene Fathers*, edited by A. Roberts and J. Donaldson, 3:503–876. Edinburgh: T. & T. Clark, 1993.
———. *Apology*. In *Ante-Nicene Fathers*, edited by A. Roberts and J. Donaldson, 3:27–116. Edinburgh: T. & T. Clark, 1993
———. *De Anima*. In *Ante-Nicene Fathers*, edited by A. Roberts and J. Donaldson, 3:335–447. Edinburgh: T. & T. Clark, 1993.
———. *De Cultu Feminarum*. In *Ante-Nicene Fathers*, edited by A. Roberts and J. Donaldson, 4:36–54. Edinburgh: T. & T. Clark, 1993.
———. *De Praescriptione*. In *Ante-Nicene Fathers*, edited by A. Roberts and J. Donaldson, 3:453–502. Edinburgh: T. & T. Clark, 1993.
Thackray, A. "Matter in a Nut-Shell: Newton's *Opticks* and Eighteenth-Century Chemistry." In *Newton*, edited by I. B. Cohen and R. S. Westfall, 87–96. New York: Norton, 1995.
Tillich, P. *Systematic Theology*. Vol. 1. London: Nisbet, 1964.

Torrance, T. F. "Arnoldshian Theses." *Scottish Journal of Theology* 15 (1962) 4–21.
———. *Divine and Contingent Order*. Edinburgh: T. & T. Clark, 1981.
———. *The Mediation of Christ*. Edinburgh: T. & T. Clark, 1995.
———. "Natural Theology in the Thought of Karl Barth." In *Transformation and Convergence in the Frame of Knowledge*, edited by T. F. Torrance, 287–94. Edinburgh: T. & T. Clark, 1995.
———. "The Problem of Natural Theology in the Thought of Karl Barth." *Religious Studies* 6 (1970) 121–35.
———. "Realism and Openness in Scientific Inquiry." *Zygon* 23, no. 2 (1988) 159–69.
———. "Science, Theology, Unity." *Scottish Journal of Theology* 21 (1964) 149–54.
———. *Space, Time and Resurrection*. Edinburgh: T. & T. Clark, 1976.
———. *Space, Time and Incarnation*. Edinburgh: T. & T. Clark, 1997.
———. *Theology in Reconstruction*. London: SCM, 1965.
Trenn, T. J. "Science, Faith and Design." *Journal of Interdisciplinary Studies* (1999) 175–86.
Warburton, W. *Thinking from A to Z*. New York: Routledge, 1996.
Warfield, B. B. *The Inspiration and Authority of the Bible*. Phillipsburg, NJ: Presbyterian and Reformed, 1948.
———. "Review of George Paulin, No Struggle for Existence: No Natural Selection. A Critical Examination of the Fundamental Principles of the Darwinian Theory." In *B B Warfield: Evolution, Science and Scripture, Selected Writings*, edited by M. A. Noll and D. N. Livingstone, 252–56. Grand Rapids: Eerdmans, 2000.
Waszink, J. H. *Tertullian's De Anima*. Amsterdam: North-Holland, 1947.
Watson, F. "Hermeneutics and the Doctrine of Scripture: Why They Need Each Other." Paper presented at the Australasian Theological Forum Conference "Hermeneutics and the Authority of Scripture," Canberra, Australia, November 23–26, 2007.
Watson, G. "A Study in St Anselm's Soteriology and Karl Barth's Theological Method." *Scottish Journal of Theology* 42 (1995) 493–512.
Watts, F. "Are Science and Religion in Conflict?" *Zygon* 32 no. 1 (1997) 125–38.
———. *Science and Theology*. Aldershot, UK: Ashgate, 2002.
———. "Theology and Science." *Theology and Science* 3, no. 3 (2005) 250.
Webster, J. "Barth, Modernity and Postmodernity." In *Karl Barth: A Future for Postmodern Theology?*, edited by G. Thompson, and C. Mostert, 1–28. Adelaide: Australian Theological Forum, 2000.
———. *The Cambridge Companion to Karl Barth*. Cambridge: Cambridge University Press, 2000.
Westfall, R. S. *The Life of Isaac Newton*. Cambridge: Cambridge University Press, 1994.
———. "Newton and Alchemy." In *Occult and Scientific Mentalities in the Renaissance*, edited by B. Vickers, 315–35 Cambridge: Cambridge University Press, 1984.
———. "Newton and Christianity." In *Newton*, edited by I. B. Cohen, and R. S. Westfall, 356–70. New York: Norton, 1995.
White, A. D. *A History of the Warfare of Science with Theology in Christendom*. 2 vols. New York: Dover, 1860.
Whitehead, A. N. *Process and Reality: An Essay in Cosmology*. London: Macmillan, 1978.
Whiteside, D. "The Expanding World of Newtonian Research." *History of Science* 1, no. 1 (1962) 16–29.
Wiles, M. *Archetypal Heresy*. Oxford: Clarendon, 1996.

Williams, R. *On Christian Theology*. Oxford: Blackwell, 2000.
Wilson, E.O. *Consilience*. New ed. London: Abacus, 2006.
The Wordsworth Dictionary of Biography. Hertfordshire, UK: Helicon, 1994.
Work, Telford. "Pneumatological Relations and Christian Disunity in Theology-Science Dialogue." *Zygon* 43, no. 4 (2008) 897–908.
Worthing, M. W. *God, Creation and Contemporary Physics*. Minneapolis: Augsburg, 1996.
———. "Science and Theology—an Historical Overview." *Pacific Journal of Theology and Science* 1, no. 1 (2000) 5–11.
———. "God, Process and Cosmos: Is God Just Going Along for the Ride?" In *Interdisciplinary Perspectives on Cosmology and Biological Evolution*, edited by L. A. Trost, 153–71, Adelaide: Centre for Theology and the Natural Sciences, Berkley, 2001.
Yeo, R. "Genius, Method and Morality: Images of Newton in Britain 1760–1860." *Science in Context* 3, no. 2 (1998) 257–84.
———. *Defining Science: William Whewell, Natural Knowledge, and Public Debate in Early Victorian Britain*. Cambridge: Cambridge University Press, 1993.
Yong, Amos. "Discerning the Spirit(s) in the Natural World: Toward a Typology of 'Spirit' in the Religion and Science Conversation." *Theology and Science* 3, no. 3 (2005) 315–29.
———. "The Spirit at Work in the World: A Pentecostal-Charismatic Perspective on the Divine Action Project." *Theology and Science* 7, no. 2 (2009) 130–40.
Yu, C. "The Principle of Relativity as a Conceptual Tool in Theology." In *Science and Theology: Questions at the Interface*, edited by M. Rae, H. Regan, and J. Stenhouse, 180–210. Grand Rapids: Eerdmans, 1994.

www.ingramcontent.com/pod-product-compliance
Lightning Source LLC
Chambersburg PA
CBHW071235230426

43668CB00011B/1448